Springer Theses

Recognizing Outstanding Ph.D. Research

Aims and Scope

The series "Springer Theses" brings together a selection of the very best Ph.D. theses from around the world and across the physical sciences. Nominated and endorsed by two recognized specialists, each published volume has been selected for its scientific excellence and the high impact of its contents for the pertinent field of research. For greater accessibility to non-specialists, the published versions include an extended introduction, as well as a foreword by the student's supervisor explaining the special relevance of the work for the field. As a whole, the series will provide a valuable resource both for newcomers to the research fields described, and for other scientists seeking detailed background information on special questions. Finally, it provides an accredited documentation of the valuable contributions made by today's younger generation of scientists.

Theses are accepted into the series by invited nomination only and must fulfill all of the following criteria

- They must be written in good English.
- The topic should fall within the confines of Chemistry, Physics, Earth Sciences, Engineering and related interdisciplinary fields such as Materials, Nanoscience, Chemical Engineering, Complex Systems and Biophysics.
- The work reported in the thesis must represent a significant scientific advance.
- If the thesis includes previously published material, permission to reproduce this must be gained from the respective copyright holder.
- They must have been examined and passed during the 12 months prior to nomination.
- Each thesis should include a foreword by the supervisor outlining the significance of its content.
- The theses should have a clearly defined structure including an introduction accessible to scientists not expert in that particular field.

More information about this series at http://www.springer.com/series/8790

Alexander R. H. Smith

Detectors, Reference Frames, and Time

Doctoral Thesis accepted independently by the University of Waterloo, Canada and Macquarie University, Australia

 Springer

Alexander R. H. Smith
Department of Physics and Astronomy
Dartmouth College
Hanover, NH, USA

ISSN 2190-5053   ISSN 2190-5061   (electronic)
Springer Theses
ISBN 978-3-030-10999-8   ISBN 978-3-030-11000-0   (eBook)
https://doi.org/10.1007/978-3-030-11000-0

Library of Congress Control Number: 2019932817

This Springer imprint is published by the registered company Springer Nature Switzerland AG.
The registered company address is: Gewerbestrasse 11, 6330 Cham, Switzerland

We are like sailors who on the open sea must reconstruct their ship but are never able to start afresh from the bottom. Where a beam is taken away a new one must at once be put there, and for this the rest of the ship is used as support. In this way, by using the old beams and driftwood the ship can be shaped entirely anew, but only by gradual reconstruction.

Otto Neurath (1921)

To my parents, Randolph and Christine Smith. Your love and support has helped me in many ways.

Supervisor's Foreword

Our deepest understanding of nature today stands astride two cornerstones of twentieth-century physics: quantum mechanics and general relativity. The former has been empirically confirmed in a legion of experiments and is ubiquitous in its applications. Whether it be lasers, transistors, superfluids, molecules, condensates, magnetism, or subatomic particles, quantum physics is the essential tool required in obtaining a quantitative and predictive description of these phenomena. Likewise, general relativity has proven to be indispensable in describing our universe at large scales ranging from the solar system to galaxies to the farthest reaches of the cosmos. While experimental tests of general relativity are not as widespread, they are no less impressive in revealing subtle aspects to the fabric of reality that nobody thought possible a little more than a century ago.

Despite this impressive array of achievements, these two descriptive paradigms are at present incompatible with each other. Despite huge efforts by a very large number of theoretical physicists, we have yet to understand how to quantize general relativity—or more broadly, how to obtain a quantum description of the gravitational force. While the technical problems are formidable, the problems run much deeper than that, since each is founded on quite distinct conceptualizations of nature. General relativity posits that space and time are unified into spacetime, an entity that dynamically responds to the behaviour of matter and energy and which in turn governs motion of the same. Time and space are fungible, with different observers able to use any desired self-consistent set of coordinates to describe given phenomena, and which under the right circumstances imply such counterintuitive predictions as the expansion of the universe and the existence of black holes. Quantum physics is no less counterintuitive. It conceives the most basic properties of matter—location, momentum, angular momentum, electric flux—as having an indefinite character described by a wave function, whose ontological status is today a matter of active debate. Any given system not only can be in one of a variety of possible states but can also simultaneously be in mutually incompatible combinations of such states (such as spin-up and spin-down), whose final outcome after a given process can be predicted only in probabilistic terms. Time plays a special role, cause and effect are differently understood, and correlations between two different systems can be much stronger than non-quantum theories (such as general relativity) would allow.

It is in this context that the work Alexander Smith describes in his thesis needs to be understood. Here Alex examines some very basic foundational issues in physics connected with our understanding of space and time. Rather than following more traditional routes in string theory and loop quantum gravity, Alex explores how our very basic notions of space and time, along with our descriptions of them, are enriched by adopting new approaches and perspectives. In so doing he obtains three novel results.

First, using an operational approach based on a theoretical model of a detector with two energy levels (a simple kind of atom), he finds that such detectors can, in principle, tell us about the topological structure of space, even when this structure is hidden inside a black hole. As Alex demonstrates, this is intrinsically connected with employing a quantum detector or a pair of such detectors to probe the vacuum structure of spacetime. Quantum fluctuations of fields in this vacuum are capable of exciting a quantum detector that would otherwise remain forever in its state of lowest energy (or ground state). If a pair of detectors is present then their quantum states can become entangled, yielding a greater measure of mutual correlation than is otherwise possible. Most importantly, as Alex demonstrates, these excitation rates and entanglement properties provide a "thumbprint" of the topology of space, distinguishing between structures that would otherwise be indistinguishable to observers outside the black hole that could not exploit quantum physics.

The second result is that of obtaining the first construction of a quantum reference frame associated with translations in phase space, or shifts in the origin of coordinate axes. Comparing two such frames in a non-quantum (or classical) sense is a straightforward problem in first-year physics. However to do so in a fully quantum sense—without recourse to classical physics—remained an unsolved problem until Alex addressed it in his thesis. He made progress by generalizing the method of group twirls to the case of reference frame symmetries under non-compact groups. He was then able to use the formalism he developed to present a protocol for communication between parties not sharing a common reference frame. This work is the first step in constructing quantum reference frames for relativistic systems.

The third result is that of formulating a new notion of quantum time. In the final part of his thesis, Alex constructs a conditional interpretation of time in terms of a quantum system whose different responses to phenomena can be regarded as clock readings. Alex, for the first time, takes into account interactions between the quantum clock and the system whose time evolution it is measuring. He finds that the notion of time which emerges entails a modification to Schröedinger's equation, the foundational equation of quantum physics.

In reading this thesis you will find that Alex's investigations of each topic are well grounded, sometimes almost starting from scratch to eventually arrive at far reaching conclusions. Its common thread is that "applying the theory of quantum information to situations at the boundary of relativity and quantum theory will certainly lead to new insights into the nature of our world". This is what the thesis achieves. We are both grateful to the Natural Sciences and Engineering Research Council of Canada for making this work possible, and to both the University of Waterloo and Macquarie University where this work was carried out.

Waterloo, ON, Canada   Professor Robert B. Mann
October 2018

Acknowledgements

I have been fortunate enough to have had three incredible supervisors: Robb Mann, Marco Piani, and Daniel Terno. In your own way, each of you has shaped the way I view and practice science. You have all been tremendously supportive in everything I have done. Thank you so much.

Mehdi Ahmadi, Andrzej Dragon, Charles Dyer, Achim Kempf, Eduardo Martín-Martínez, and Edward Vrscay—I have learned much from all of you and I am grateful for the time we've spent discussing physics and mathematics. Thank you.

I wish to thank my friends and collaborators; I have had the privilege of not having to distinguish collaborators from friends: Aida Ahmadzadegan, Natacha Altamirano, Wilson Brenna, Aharon Brodutch, Eric Brown, Melanie Chanona, Aidan Chatwin-Davies, Agata Chęcińska, John Chibuk, Paulina Corona, Jen Cos, Jack Davis, Cohl Dron, Max Maaneli Derakhshani, Alexei Gilchrist, Daniel Grimmer, Thomas Guff, Laura Henderson, Jaden Hellmann, Robie Hennigar, Daniel Hirschmeier, Fern Hewitt, Maria Kovaleva, Jeffery Larkin, Adam Lewis, Dmitry Linkov, Krzysztof Lorek, Eleanor McGrath, Paul McGrath, Nick Menicucci, Sean McKay, Kavan Modi, Keith Motes, Keith Ng, Ehsan Pasha, Miok Park, Joel Percy, Claire Pilaprat, Jacques Pienaar, Roberto Pierini, Felix Pollock, Allison Sachs, Jaspreet Sahota, Maurine Seto, Fil Simovic, Renford Smith, Elizabeth Snyder, Carl Snyder, Rob Thomas, Zehua Tian, Erickson Tjoa, and Jialin Zhang. I am grateful for the conversations and memories we've shared. You have made the time spent pursuing my PhD the most rewarding of my life.

To my parents, Randolph and Christine Smith, I cannot thank you enough for your encouragement and support. You have given me the opportunity to study physics. This thesis is dedicated to you.

Thank you to my loving partner Hilary Snyder for listening to me babble about clocks, the quantum, and other silly things. You have filled my life with fun and happiness.

Contents

Parts of this thesis have been published in the following journal articles:

1. Alexander R. H. Smith, *Communicating without shared reference frames*, Physical Review A 99, 052315 (2019). DOI: 10.1103/PhysRevA.99.052315
2. Laura J. Henderson, Robie A. Hennigar, Robert B. Mann, Alexander R. H. Smith, and Jialin Zhang, *Harvesting Entanglement from the Black Hole Vacuum*, Classical and Quantum Gravity 35, 21LT02 (2018). DOI: 10.1088/1361-6382/aae27e
3. Alexander R. H. Smith, Marco Piani, and Robert B. Mann, *Quantum reference frames associated with noncompact groups: the case of translations and boosts, and the role of mass*, Physical Review A 94, 012333 (2016). DOI: 10.1103/PhysRevA.94.012333
4. Eduardo Martín-Martínez, Alexander R. H. Smith, and Daniel R. Terno, *Spacetime structure and vacuum entanglement*, Physical Review D 93, 044001 (2016). DOI: 10.1103/PhysRevD.93.044001
5. Alexander R. H. Smith and Robert B. Mann, *Looking Inside a Black Hole*, Classical and Quantum Gravity 31, 082001 (2014). DOI: 10.1088/0264-9381/31/8/082001

Parts of this thesis will appear in the following forthcoming articles:

1. Alexander R. H. Smith and Mehdi Ahmadi, *Quantizing time: Interacting clocks and systems*, arXiv:1712.00081 [quant-ph]

Chapter 1
Introduction

At the beginning of his book *Quantum Field Theory in Curved Spacetime and Black Hole Thermodynamics* [3], Robert Wald outlines the main applications of quantum field theory on curved spacetime:

> Its two applications of greatest interest are to phenomena occurring in the very early universe and to phenomena occurring in the vicinity of black holes. During the past twenty-five years such phenomena have been explored theoretically, and some unexpected and intriguing results have been obtained. Most prominent among these was the discovery by Hawking that particle creation occurs in the vicinity of black holes. As a direct consequence, a deep connection was obtained between the laws of black hole physics and the ordinary laws of thermodynamics. The Hawking effect and its implications are probably the most valuable clues we have, at present, as to the fundamental features that a quantum theory of gravity is likely to possess.

Since the publication of Wald's book in 1994, physics has taken on an increasingly information-theoretic flavour. This has led to a quantum theory of information which promises computers with unprecedented power and unconditionally secure cryptography, and perhaps most importantly, it has offered us a new perspective on quantum theory itself. The aim of this thesis is to use the large toolbox of quantum information theory to study problems whose solution requires both quantum mechanics and relativity, with the hope that these investigations may offer new *unexpected and intriguing* results about possible features of quantum gravity.

This thesis is broken into three distinct parts, each of which is self-contained; these parts may be read in any order. We give a brief description of each chapter here.

© Springer Nature Switzerland AG 2019

A. R. H. Smith, *Detectors, Reference Frames, and Time*, Springer Theses,

https://doi.org/10.1007/978-3-030-11000-0_1

1.1   Part I: Detectors in Curved Spacetimes

Chapter 2: Quantum Field Theory on Curved Spacetimes
We begin by describing the quantization of a real scalar field on a curved spacetime; we limit ourselves to the study of such fields because they are mathematically simple while exhibiting a range of quantum field theory effects on curved spacetimes. Beginning with the action for a real scalar field, we demonstrate how the equations of motion for the field come about and introduce an appropriate inner product on the space of solutions. We then describe the canonical quantization of this field theory and emphasize the non-uniqueness of the vacuum state. We describe in detail the particle interpretation of such a field theory with an emphasis on the importance of an operational definition of particles.

Chapter 3: The Unruh-DeWitt Detector and Entanglement Harvesting
In this chapter we introduce the Unruh-DeWitt detector: a two-level quantum system interacting locally with a quantum field moving through spacetime along a timelike trajectory. We present a physical motivation for this detector model [2]. Then we derive the transition probability that the detector beginning in its ground state transitions to its excited state to leading order in the interaction strength between the detector and field. For detectors in curved spacetimes, we express this transition probability in terms of the Wightman function of the field. We also introduce the transition rate of such a detector and express it in terms of the Wightman function.

We then study the entanglement harvesting protocol in which two detectors begin in a separable state and as a result of their local interaction with the field become entangled. This protocol is generalized to detectors in arbitrary curved spacetimes admitting a Wightman function, and the final state of the two detectors is derived to all orders in perturbation theory; to leading order in the interaction strength, the final state of these detectors is expressed in terms of the Wightman function. We then compute several measures of entanglement and correlations (negativity, concurrence, entanglement of formation, and correlations between local measurements of the detectors) quantifying the amount of entanglement that results between the two detectors.

The measurement process involves an interaction between a measuring apparatus and the system to be measured. Viewing a collection of detectors as a measuring apparatus and a quantum field on curved spacetime as the system to be measured, we derive the associated measurement model and identify the observables on the field Hilbert space these detectors measure; the leading order contribution to the POVM elements defining these observables is given. We use this measurement model framework to rederive the transition probability of a single detector. We comment on possible applications of these results.

Chapter 4: Unruh-DeWitt Detectors in Quotients of Minkowski Space
We apply the entanglement harvesting protocol developed in Chap. 3 to detectors in Minkowski space and two cylindrical spacetimes constructed by topological identifications of Minkowski space. To do so, we review the image sum derivation

of the Wightman functions in these cylindrical spacetimes. We investigate how the entanglement harvesting protocol depends on the global topology of these spacetimes.

Chapter 5: Unruh-DeWitt Detectors in (2+1)-Dimensional Black Hole Spacetimes

In this chapter we apply the formalism developed in Chap. 3 to detectors in two black hole spacetimes: the BTZ black hole and the $\mathbb{R}P^2$ geon. These spacetimes are constructed by topological identifications of the (2+1)-anti-de Sitter space. Using this fact, we review the image sum derivation of the Wightman function in both spacetimes. The transition rate of detectors operating in the exterior region of both spacetimes is evaluated.

We will see that the $\mathbb{R}P^2$ geon black hole is an intermediate between a stationary and dynamical spacetime, in the sense that the non-stationary features are hidden behind its horizons. We demonstrate that a detector in the exterior region of the $\mathbb{R}P^2$ geon spacetime is sensitive to these features. We will show that this detector develops a time-dependent transition rate which oscillates around the transition rate of an identical detector in the BTZ spacetime. This is surprising seeing as the BTZ black hole and $\mathbb{R}P^2$ geon are locally indistinguishable from one another in the region in which the detectors are operating.

Next we realize the entanglement harvesting protocol for two detectors located in the exterior region of the BTZ black hole. This allows us to investigate operationally the entanglement structure of the Hartle-Hawking vacuum as seen by two detectors, in particular, how this entanglement depends on the detectors proximity to the black hole horizon. This is the first example of the entanglement harvesting protocol in a black hole spacetime.

1.2   Part II: Quantum Reference Frames

Chapter 6: Quantum Reference Frames Associated with Noncompact Groups

The language used to describe a collection of reference frames is group theory. This is because changes of a reference form a group. In this chapter, we explore the consequences of replacing a classical reference frame with a quantum one. We find that the relational description of a quantum state suitable for a system defined with respect to a reference associated with a compact group does not generalize to the case of noncompact groups. This is a result of the group average over a noncompact group producing non-normalizable states; this average is known as the G-twirl. However, for classical reference frames associated with the groups of spatial translations and inertial reference frames, we show that the G-twirl over these noncompact groups singles out a relational state which corresponds to tracing out degrees of freedom associated with the centre-of-mass of a composite system.

We then examine the informational properties of this relational description for quantum systems of two and three particles prepared in fully separable Gaussian states with respect to an external reference frame. In particular we quantify the entanglement that appears between the centre-of-mass and relational degrees of freedom of these systems, as well as the entanglement that appears among the relational degrees of freedom. We investigate how this entanglement depends on the mass of the particles and their state with respect to an external frame.

Motivating this investigation is the aspiration for a relational quantum theory, which will certainly require a description of quantum systems with respect to other quantum systems.

Chapter 7: Communication Without a Shared Reference Frame
We generalize a communication protocol introduced by Bartlett et al. [1] in which two parties communicating do not share a classical reference frame, to the case when changes of their reference frames form a one-dimensional noncompact Lie group. Alice sends to Bob the state $\rho_R \otimes \rho_S$, where ρ_S is the state of the system Alice wishes to communicate and ρ_R is a token of her reference frame. Because Bob is ignorant of the relationship between his reference frame and Alice's, he will describe the state $\rho_R \otimes \rho_S$ as an average over all possible reference frames. Bob measures the reference token and applies a correction to the system Alice wished to communicate conditioned on the outcome of the measurement. The recovered state ρ'_S is decohered with respect to ρ_S, the amount of decoherence depending on the properties of the reference token ρ_R.

We present an example of this protocol when Alice and Bob do not share a reference frame associated with the one-dimensional translation group and use the fidelity between ρ_S and ρ'_S to quantify the success of the recovery operation.

1.3   Part III: Quantizing Time

Chapter 8: The Conditional Probability Interpretation of Time: The Case of Interacting Clocks
The conditional probability interpretation of time is based upon conditioning a solution to the Wheeler-DeWitt equation on a subsystem of the universe being in a state corresponding to the time t; this subsystem serves as a clock. This procedure assigns a conditional state to the rest of the universe $|\psi_S(t)\rangle$, which satisfies the Schrödinger equation.

We generalize the conditional probability interpretation of time to take into account an interaction between the clock and system comprising the rest of the universe. Heuristically, we should expect such a coupling between any clock and system at some scale resulting from their gravitational interaction. We find that the conditional state $|\psi_S(t)\rangle$ satisfies a time-nonlocal modified Schrödinger equation in which the system Hamiltonian is replaced with a self-adjoint integral operator that depends on the interaction between the clock and the rest of the universe. A series solution to the modified Schrödinger equation is constructed analogous to the Dyson series.

1.4   Notation and Conventions

- In Chaps. 2–4 the signature of the metric is chosen to be mostly negative $+---$, while in Chap. 5 the signature is mostly positive $-+++$. This is done so that the results presented in each chapter are easily comparable to existing literature.
- Natural units are used throughout $\hbar = c = 1$.
- Here is a list of the symbols used and their names.

$\mathcal{H}, \mathcal{K}$	Hilbert space
$\mathcal{H}_D, \mathcal{H}_A, \mathcal{H}_B$	Detector Hilbert space
$\mathcal{H}_\phi$	Scalar field Hilbert space
$\mathcal{H}_C$	Clock Hilbert space
$\mathcal{H}_S$	System Hilbert space
$\mathcal{H}_{kin}$	Kinematical Hilbert space $\mathcal{H}_C \otimes \mathcal{H}_S$
$\mathcal{H}_{ph}$	Physical Hilbert space
$\mathcal{L}(\mathcal{H})$	The space of bounded linear operators on $\mathcal{H}$
$\mathcal{L}_s(\mathcal{H})$	The space of bounded self-adjoint operators on $\mathcal{H}$
$\mathcal{S}(\mathcal{H})$	The space of states on $\mathcal{H}$
$\mathcal{U}(\mathcal{H})$	The space of unitary operators on $\mathcal{H}$
$\mathcal{E}(\mathcal{H})$	The space of positive operator valued measures (POVM) on $\mathcal{H}$
G	A group
$e \in G$	The identity element of the group G
T_n	n-dimensional translation group
Z	The group of integers
Z_2	The cyclic group of dimension 2
ℓ	AdS length scale

References

1. S.D. Bartlett, T. Rudolph, R.W. Spekkens, P.S. Turner, Quantum communication using a bounded-size quantum reference frame. New J. Phys. **11**, 063013 (2009)
2. A. Pozas-Kerstjens, E. Martín-Martínez, Entanglement harvesting from the electromagnetic vacuum with hydrogenlike atoms. Phys. Rev. D **94**, 064074 (2016)
3. R.M. Wald, *Quantum Field Theory in Curved Spacetime and Black Hole Thermodynamics* (The University of Chicago Press, Chicago, 1994)

Part I
Detectors in Curved Spacetimes

Chapter 2
Quantum Field Theory on Curved Spacetimes

Quantum field theory on curved spacetimes examines the behaviour of quantum fields in the presence of a classical gravitational field described by the theory of general relativity. It is used to study phenomena where both gravity and the quantum nature of the field are important, but the quantum properties of the gravitational field itself can be ignored. Scalar field theory on curved spacetime exhibits many novel features of quantum field theory on curved spacetimes with minimal mathematical complexity. For this reason, we will consider scalar field theory exclusively, while acknowledging many of the results presented can be generalized to higher spin fields. The treatment presented here follows [1, 4].

The action describing a real scalar field $\phi(x)$ with mass m on a n-dimensional curved spacetime is

$$S = \int dx^n \, \mathcal{L}(x) = \int dx^n \, \frac{1}{2}\sqrt{-g}\left(\partial_\mu \phi(x)\, \partial^\mu \phi(x) - \left[m^2 + \xi R(x)\right]\phi(x)^2\right),$$

(2.1)

where g is the determinant of the metric $g_{\mu\nu}$ of the spacetime in which the field lives. The scalar field is coupled to the gravitational field by the presence of the term $\xi R(x)$ in the action, where ξ is a numerical factor and $R(x)$ is the Ricci scalar, which is the trace of the Ricci curvature tensor

$$R_{\alpha\beta} := \partial_\rho \Gamma^\rho{}_{\beta\alpha} - \partial_\beta \Gamma^\rho{}_{\rho\alpha} + \Gamma^\rho{}_{\rho\lambda}\Gamma^\lambda{}_{\beta\alpha} - \Gamma^\rho{}_{\beta\lambda}\Gamma^\lambda{}_{\rho\alpha},$$

(2.2)

where $\Gamma^\lambda{}_{\mu\nu}$ denote the Christoffel symbols associated with the metric $g_{\mu\nu}$

$$\Gamma^\lambda{}_{\mu\nu} := \frac{1}{2}g^{\lambda\rho}\left(\partial_\nu g_{\rho\mu} + \partial_\mu g_{\rho\nu} - \partial_\rho g_{\mu\nu}\right).$$

(2.3)

Extremizing the action S with respect to $\phi(x)$ yields the equation of motion which is satisfied by the field

© Springer Nature Switzerland AG 2019
A. R. H. Smith, *Detectors, Reference Frames, and Time*, Springer Theses,
https://doi.org/10.1007/978-3-030-11000-0_2

$$[\Box + m^2 + \xi R(x)]\phi(x) = 0, \tag{2.4}$$

where $\Box \phi(x) := (-g)^{-1/2} \partial_\mu [(-g)^{1/2} g^{\mu\nu} \partial_\nu \phi(x)]$.

There exists a set of mode functions $u_i(x)$, where i indicates all the quantities required to label the mode, that form a complete orthonormal set of solutions to Eq. (2.4) with respect to the inner product

$$(\phi_1, \phi_2) = - \int_\Sigma \frac{d\Sigma^\mu}{\sqrt{-g}} \left(\phi_1(x) \partial_\mu \phi_2(x)^* - [\partial_\mu \phi_1(x)] \phi_2(x)^* \right), \tag{2.5}$$

where $d\Sigma^\mu = n^\mu d\Sigma$, n^μ is a future-pointing unit vector orthogonal to the spacelike Cauchy hypersurface Σ, $d\Sigma := \det h \, dx^{n-1}$ is the invariant measure on Σ, and h the induced metric on Σ. Explicitly, the orthonormality relationships between the mode functions are[1]

$$(u_i, u_j) = \delta_{ij}, \quad (u_i^*, u_j^*) = -\delta_{ij}, \quad \text{and} \quad (u_i, u_j^*) = 0. \tag{2.6}$$

Suppose the spacetime admits a timelike Killing vector ξ_t, which generates time translations in the coordinate t. The remaining $n - 1$ coordinates will be denoted by bold letters, that is, $x = (t, \mathbf{x})$. In such spacetimes, there exists a set of mode functions which are eigenfunctions of the operator $i\partial_t$

$$i\partial_t u_k(x) = \omega_k u_k(x), \tag{2.7}$$

where $\omega_i \geq 0$. The field $\phi(x)$ may be expanded in terms of these modes as

$$\phi(x) = \sum_k \left[u_k(x) a_k + u_k^*(x) a_k^\dagger \right]. \tag{2.8}$$

Quantization proceeds by promoting the field $\phi(x)$ and its conjugate momentum $\pi(x) := \partial \mathcal{L} / \partial [\partial_t \phi(x)]$, where $\mathcal{L}$ is the Lagrangian density appearing in Eq. (2.1), to operators and imposing the canonical commutation relations

$$[\phi(t, \mathbf{x}), \pi(t, \mathbf{y})] = i\delta^{n-1}(\mathbf{x} - \mathbf{y}), \tag{2.9}$$

and

$$[\phi(t, \mathbf{x}), \phi(t, \mathbf{y})] = [\pi(t, \mathbf{x}), \pi(t, \mathbf{y})] = 0. \tag{2.10}$$

---

[1] In this chapter we will confine the solutions $u_i(x)$ to a $(n-1)$-tours of side length L, i.e. box normalized solutions. This results in the solutions being labelled by discrete indices. To convert to the continuum normalization one should replace $\left(\frac{2\pi}{L}\right)^{n-1} \sum_i$ with $\int dk^{n-1}$ and the Kronecker delta with the Dirac delta function.

These commutation relations imply the following commutation relations for the operators a_k and $a_k^\dagger$

$$[a_i, a_j^\dagger] = \delta_{ij} \quad \text{and} \quad [a_i, a_j] = [a_i^\dagger, a_j^\dagger] = 0. \tag{2.11}$$

The Hamiltonian operator associated with the scalar field may be constructed from the classical energy-momentum tensor

$$T_{\mu\nu} = \frac{2}{\sqrt{-g}} \frac{\delta S}{\delta g^{\mu\nu}}, \tag{2.12}$$

by integrating over a constant time spacelike hypersurface Σ

$$H := \int_\Sigma d\Sigma^\mu \, \xi_t^\nu \, T_{\mu\nu} = \sum_i \omega_k \left(a_k^\dagger a_k + 1/2 \right). \tag{2.13}$$

States of the field span a Hilbert space, a convenient basis for which is known as the Fock basis. The construction of the Fock basis begins by defining the vacuum state $|0\rangle$, which has the property that it is the state with lowest energy as measured by the operator H in Eq. (2.13). As a consequence, the vacuum state may equivalently be defined as the state that it is annihilated by all the operators a_k

$$a_k |0\rangle = 0 \quad \forall k. \tag{2.14}$$

The Fock basis states are constructed by repeatedly acting on the vacuum state with the operators $a_k^\dagger$

$$\left| n_{k_1}, n_{k_2}, \dots, n_{k_j} \right\rangle = \frac{1}{\sqrt{n_{k_1}! \, n_{k_2}! \, \dots \, n_{k_j}!}} \left(a_{k_1}^\dagger \right)^{n_{k_1}} \left(a_{k_2}^\dagger \right)^{n_{k_2}} \dots \left(a_{k_j}^\dagger \right)^{n_{k_j}} |0\rangle. \tag{2.15}$$

Now consider the operator

$$N_k := a_k^\dagger a_k. \tag{2.16}$$

The eigenstates of N_k are the Fock basis states given in Eq. (2.15)

$$N_{k_i} \left| n_{k_1}, n_{k_2}, \dots, n_{k_j} \right\rangle = n_{k_i} \left| n_{k_1}, n_{k_2}, \dots, n_{k_j} \right\rangle, \tag{2.17}$$

with eigenvalue n_{k_i} corresponding to the integer labelling the number of times the operator $a_{k_i}^\dagger$ has to be applied to the vacuum $|0\rangle$ to create the state $\left| n_{k_1}, n_{k_2}, \dots, n_{k_j} \right\rangle$. From the definitions of N_k and H above, it is seen that

$$[N_k, H] = 0, \tag{2.18}$$

which implies that eigenstates of N_k are eigenstates of H. Furthermore, for each increment of n_{k_i} defining a particular Fock basis state, the expectation value of H increments by ω_{k_i}. This observation suggests we interpret n_{k_i} as labelling the number of quanta or particles[2] with energy ω_{k_i} in the field state given in Eq. (2.17). The total number of particles in a given state is then counted by the number operator

$$N := \sum_k N_k. \tag{2.19}$$

For example, the number of particles in a Fock basis state is the eigenvalue of N

$$N \left| n_{k_1}, n_{k_2}, \ldots, n_{k_j} \right\rangle = \left(\sum_i n_{k_i} \right) \left| n_{k_1}, n_{k_2}, \ldots, n_{k_j} \right\rangle, \tag{2.20}$$

which is seen to be $\sum_i n_{k_i}$.

Having introduced the Fock basis states and the notion of particles, a physical interpretation is now available for the operators a_k and $a_k^\dagger$. From Eqs. (2.15) and (2.17) we see that the operator a_k reduces and the operator $a_k^\dagger$ increases the number of particles in the mode labelled by k. We will therefore refer to a_k and $a_k^\dagger$ as creation and annihilation operators.

The quantization procedure presented above depends on our choice to expand the field in terms of the mode functions $u_k(x)$ in Eq. (2.8) and imposing the canonical commutation relations between the field operator $\phi(x)$ and its canonical momentum $\pi(x) := \partial\mathcal{L}/\partial[\partial_t\phi(x)]$, which depends on a time derivative. Consequently, this quantization procedure depends on the choice of coordinate system.

In Minkowski space a preferred set of modes, and the vacuum they define, is singled out by the requirement that the vacuum state is the same for all inertial observers. The ability to do this depends on the isometry group of the spacetime, which in Minkowski space is the Poincaré group. However, a curved spacetime may not admit an isometry group and thus, in general, there is no preferred coordinate system to carry out the quantization procedure with and consequently no preferred vacuum state exists.

With this in mind, let us consider another complete set of orthonormal mode function $v_j(x)$ which solve Eq. (2.4). The field may be expanded in terms of these mode functions as

$$\phi(x) = \sum_j \left[v_j(x)b_j + v_j^*(x)b_j^\dagger \right]. \tag{2.21}$$

---

[2]The use of the word "particle" here is different than how the word is used in everyday language. The word particle commonly refers to an object with a well-defined energy, momentum, and position. Here, the use of the word particle refers to excitations of the field that have a well-defined energy. However, these particles are global excitations of the field and therefore do not have a well-defined position.

Quantization proceeds as before: b_j and $b_j^\dagger$ are promoted to operators satisfying the commutation relations

$$[b_i, b_j^\dagger] = \delta_{ij} \quad \text{and} \quad [b_i, b_j] = [b_i^\dagger, b_j^\dagger] = 0. \tag{2.22}$$

The decomposition of the field in Eq. (2.21) defines a different vacuum state $|\bar{0}\rangle$, which is annihilated by every annihilation operator b_j

$$b_j |\bar{0}\rangle = 0 \quad \forall j. \tag{2.23}$$

This in turn defines a new Fock basis and a new notion of particle.

As both sets of modes, $u_i(x)$ and $v_j(x)$, are complete, they may be expanded in terms of one another as

$$v_j(x) = \sum_i \left[\alpha_{ji} u_i(x) + \beta_{ji} u_i^*(x)\right], \tag{2.24a}$$

$$u_i(x) = \sum_j \left[\alpha_{ji}^* v_j(x) - \beta_{ji} v_j^*(x)\right]. \tag{2.24b}$$

This transformation between different sets of mode functions is known as a Bogoliubov transformation and the coefficients α_{ij} and β_{ij} as the Bogoliubov coefficients. The Bogoliubov coefficients are extracted from the above relations by taking appropriate inner products and using the orthonormality of the mode functions, with the result

$$\alpha_{ij} = (v_i, u_j) \quad \text{and} \quad \beta_{ij} = -(v_i, u_j^*). \tag{2.25}$$

Equating the field expansion in terms of $u_i(x)$ in Eq. (2.8) with the field expansion in terms of $v_j(x)$ in Eq. (2.21), and employing the relations given in Eq. (2.24), allows us to express the operators b_j in terms of the operators a_i and $a_i^\dagger$ and vice versa

$$a_i = \sum_j \left[\alpha_{ji} b_j + \beta_{ji} b_j^\dagger\right], \tag{2.26a}$$

$$b_j = \sum_i \left[\alpha_{ji}^* a_i - \beta_{ji}^* a_i^\dagger\right]. \tag{2.26b}$$

From the above relations we see that if $\beta_{ij} \neq 0$, the vacuum state $|0\rangle$ defined by the mode functions $u_i(x)$ will contain particles as counted by the number operator $\bar{N} := \sum_j b_j^\dagger b_j$ associated with the mode functions $v_j(x)$,

$$\langle 0| \, \bar{N} \, |0\rangle = \sum_j |\beta_{ij}|^2 \, . \tag{2.27}$$

The same is true of the number of particles counted by the operator N associated with the modes $u_i(x)$ given the state was in the vacuum $|\bar{0}\rangle$ associated with the $v_j(x)$ modes,

$$\langle \bar{0}| \, N \, |\bar{0}\rangle = \sum_j |\beta_{ij}|^2 \, . \tag{2.28}$$

From the preceding discussion it is clear that there is no unique vacuum state of the field or notion of particle. The vacuum state depends on which set of mode functions are used to expand the field and carry out the quantization procedure with. The question then arises, is there a preferred set of mode functions to quantize the field and define particles and the vacuum with, which matches our experience of particles and no particles?

As posed, the question is unanswerable. Our experience of particles necessarily requires us to specify the measurement process by which we detect particles. This includes the specification of the interaction between the field and the employed particle detector. This leads to an operational definition of a particle: a particle is what is detected by a particle detector.

These conclusions are true even in Minkowski space. As discussed above, the conventional vacuum state is defined as the state for which no inertial detector registers particles. In other words, the vacuum state is invariant under the Poincaré group. Indeed, an observer accelerating through this vacuum will see a thermal bath of particles at a temperature proportional to their acceleration; this is the well-known Unruh effect [2, 3, 5]. An analogous construction of a vacuum state may be available in other highly symmetric spacetimes. However, in an arbitrary spacetime lacking an isometry group, freely falling observers (the curved space generalization of inertial observers) will not agree on a vacuum state. This further motivates the operational definition of a particle.

In the next chapter we will introduce a specific model of a particle detector, the Unruh-DeWitt detector, and use this model to study vacuum excitations, field entanglement, and relativistic measurements of quantum fields in an operational manner.

References

1. N.D. Birrell, P.C.W. Davies, *Quantum Fields in Curved Space* (Cambridge University Press, Cambridge, 1982)
2. P.C.W. Davies, Scalar production in Schwarzschild and Rindler metrics. J. Phys. A Gen. Phys. **8**, 609 (1975)

3. S.A. Fulling, Nonuniqueness of canonical field quantization in Riemannian space-time. Phys. Rev. D **7**, 2850 (1973)
4. V.F. Mukhanov, S. Winitzki, *Introduction to Quantum Effects in Gravity* (Cambridge University Press, Cambridge, 2007)
5. W.G. Unruh, Notes on blackhole evaporation. Phys. Rev. D **14**, 870 (1976)

Chapter 3
The Unruh-DeWitt Detector
and Entanglement Harvesting

This chapter begins by introducing the Unruh-DeWitt detector in Sect. 3.1. In Sect. 3.2 we apply this detector model to the phenomenon of entanglement harvesting in curved spacetimes that admit a Wightman function. In Sect. 3.3 we identify the field observables that a collection of Unruh-DeWitt detectors measure. We conclude this chapter in Sect. 3.4 with a summary of the results presented.

3.1 The Unruh-DeWitt Detector

As discussed at the end of Chap. 2, the definition of a particle is ultimately an operational one that requires the specification of the measuring process used to detect particles. This requires the specification of the interaction of a particle detector with a field. This definition is not unique—there are as many definitions of particle as there are particle detectors. In this section, we describe a commonly used measuring process known as the Unruh-DeWitt detector [15, 48].

We seek a measuring process that is physically relevant and mathematically simple. To this end, consider as a measuring apparatus a two-level atom. Let $|0\rangle_D$ and $|1\rangle_D$ denote the ground state and excited state of the atom (detector), which are separated by an energy gap Ω. These states form an orthonormal basis for the Hilbert space $\mathcal{H}_D \simeq \mathbb{C}^2$ associated with internal degrees of freedom of the atom, whose free evolution is governed by the Hamiltonian

$$H_0 = \frac{\Omega}{2} \big(|1\rangle_D\langle 1|_D - |0\rangle_D\langle 0|_D \big). \tag{3.1}$$

The atom's interaction with an electromagnetic field is described by quantum electrodynamics. For a two-level atom this interaction is well approximated by the dipole interaction Hamiltonian

$$H_{int} = e\mathbf{X} \cdot \mathbf{E}, \tag{3.2}$$

© Springer Nature Switzerland AG 2019
A. R. H. Smith, *Detectors, Reference Frames, and Time*, Springer Theses,
https://doi.org/10.1007/978-3-030-11000-0_3

where $\mathbf{E}$ is the electric field and $e\mathbf{X}$ is the dipole operator, with e being the charge of the dipole and $\mathbf{X} = \int d\mathbf{x}\,\mathbf{x}\,|\mathbf{x}\rangle\langle\mathbf{x}|$ its position operator. This interaction Hamiltonian is widely used in the field of quantum optics to model the light-matter interaction [25, 40]. Expanding H_{int} in terms of the excited and ground states of the atom and moving to the interaction picture[1] yield

$$H_I = e^{i\Omega\tau}\,|1\rangle_D\langle 0|_D \otimes \int d\mathbf{x}\,\mathbf{F}(\mathbf{x})\cdot\mathbf{E}(t,\mathbf{x}) + e^{-i\Omega\tau}\,|0\rangle_D\langle 1|_D \otimes \int d\mathbf{x}\,\mathbf{F}^*(\mathbf{x})\cdot\mathbf{E}(t,\mathbf{x}),$$

(3.3)

where $\mathbf{F}(\mathbf{x}) := \psi_1^*(\mathbf{x})\,\mathbf{x}\,\psi_0(\mathbf{x})$ and $\psi_0(\mathbf{x}) := \langle\mathbf{x}|0\rangle$ and $\psi_1(\mathbf{x}) := \langle\mathbf{x}|1\rangle$ are the ground state and excited state wave functions in the position basis.[2]

From Eq. (3.3), we see that the function $\mathbf{F}(\mathbf{x})$ acts as a smearing function with support localized around the atom. We will make the assumption that the spatial extent of the atom is sufficiently localized so that the smearing function may be approximated as a delta function $\mathbf{F}(\mathbf{x}) \approx \delta^3(\mathbf{x} - \mathbf{x}_D(\tau))$, where $x_D(\tau) := \{t_D(\tau), \mathbf{x}_D(\tau)\}$ is the trajectory of the atom parametrized in terms of its proper time τ. This models the atom as coupling to the electric field at a point, which is a good assumption so long as the spatial extent of the atom is negligible as compared to the wavelength of the radiation that is resonant with the atom's energy gap [2].

To arrive at the Unruh-DeWitt interaction Hamiltonian, in Eq. (3.3) we replace the electric field with a scalar field $\phi(x)$ and introduce a switching function $\chi_D(\tau) \in [0, 1]$, which controls the duration of the interaction between the atom and field. Doing so yields the Unruh-DeWitt interaction Hamiltonian

$$H_D(\tau) = \lambda\chi_D(\tau)\left(\sigma^+(\tau) + \sigma^-(\tau)\right) \otimes \phi[x_D(\tau)],$$

(3.4)

where $\lambda \ll 1$ is the interaction strength and we have defined the ladder operators

$$\sigma^+(\tau) := e^{i\Omega\tau}\,|1\rangle_D\langle 0|_D \quad \text{and} \quad \sigma^-(\tau) := e^{-i\Omega\tau}\,|0\rangle_D\langle 1|_D.$$

(3.5)

When the interaction of a two-level atom with the electromagnetic field is approximated in this way, the atom is referred to as an Unruh-DeWitt detector or simply detector. While the Unruh-DeWitt interaction Hamilton is unable to capture the vector nature of the electric field, it has been shown that it approximates well the light-matter interaction in other regards [2, 27]. For a detailed comparison of different light-matter interaction models within this context see [33].

The time evolution of the detector and field during the measuring process, that is, when the detector and field are interacting, is described by the unitary operator generated by the interaction Hamiltonian in Eq. (3.4)

---

[1] In the interaction picture, the interaction Hamiltonian is $H_I := e^{iH_0 t}H_{int}e^{-iH_0 t}$.

[2] The diagonal elements of H_I vanish because $\langle 0|\mathbf{X}|0\rangle = \langle 1|\mathbf{X}|1\rangle = 0$, which comes from the fact that $\psi_0(\mathbf{x})$ and $\psi_1(\mathbf{x})$ are symmetric around $\mathbf{x} = 0$ due to the Coulomb interaction between the nucleus and electron of the atom being symmetric around $\mathbf{x} = 0$.

$$U := \mathcal{T} \exp\left[-i \int d\tau\, H_D(\tau)\right]$$

$$= 1 + (-i) \int d\tau\, H_D(\tau) + \frac{(-i)^2}{2} \int d\tau d\tau'\, \mathcal{T} H_D(\tau)\, H_D(\tau') + \mathcal{O}\left(\lambda^3\right),$$
(3.6)

where the integration is over the interval $\tau \in (-\infty, \infty)$ and $\mathcal{T}$ is the time ordering operator defined as $\mathcal{T} A(t) B(t') := \theta(t - t') A(t) B(t') + \theta(t' - t) B(t') A(t)$.

Suppose that prior to the measuring process ($\tau \to -\infty$) the detector is prepared in the ground state $|0\rangle_D$ and the field in an appropriately defined vacuum state $|0\rangle$. After the measuring process, the final state of the field and detector will be

$$U\, |0\rangle_D\, |0\rangle$$

$$= \left(1 - i \int d\tau\, H_D(\tau) - \frac{1}{2} \int d\tau d\tau'\, \mathcal{T} H_D(\tau) H_D(\tau') + \mathcal{O}\left(\lambda^3\right)\right) |0\rangle_D\, |0\rangle$$

$$= |0\rangle_D\, |0\rangle - i\lambda \int d\tau\, \chi_D(\tau) e^{i\Omega_D \tau}\, |1\rangle_D \otimes \phi[x_D(\tau)]\, |0\rangle$$

$$- \frac{\lambda^2}{2} \int d\tau d\tau'\, \chi_D(\tau)\chi_D(\tau') \mathcal{T}\left[\sigma^-(\tau)\sigma^+(\tau') + \sigma^+(\tau)\sigma^-(\tau')\right] |0\rangle_D$$

$$\otimes \mathcal{T}\phi[x_D(\tau)]\, \phi[x_D(\tau')]\, |0\rangle + \mathcal{O}\left(\lambda^3\right).$$
(3.7)

The final state of the detector alone, $\rho_D \in \mathcal{S}(\mathcal{H}_D)$, is obtained by tracing over the field degrees of freedom in Eq. (3.7)

$$\rho_D := \mathrm{tr}_\phi\left[U\left(|0\rangle\langle 0|_D \otimes |0\rangle\langle 0|\right)U^\dagger\right] = \begin{pmatrix} 1 - P_D & 0 \\ 0 & P_D \end{pmatrix} + \mathcal{O}\left(\lambda^4\right),$$
(3.8)

where in the last equality we expressed the detector density matrix in the basis $\{|0\rangle_D, |1\rangle_D\}$. P_D is the probability that the detector has transitioned to its excited state

$$P_D := \lambda^2 \int d\tau d\tau'\, \chi_D(\tau)\chi_D(\tau') e^{-i\Omega(\tau - \tau')} W\left(x_D(\tau), x_D(\tau')\right),$$
(3.9)

and $W(x, x')$ is the vacuum Wightman function

$$W(x, x') := \langle 0|\, \phi(x)\phi(x')\, |0\rangle.$$
(3.10)

As the field $\phi(x)$ satisfies the wave equation in Eq. (2.4), so too does the Wightman function

$$\left[\Box_x + m^2 + \xi R(x)\right] W(x, x') = 0,$$
(3.11)

where the subscript x on $\Box_x$ denotes that the derivatives appearing in the definition of the d'Alembertian below Eq. (2.4) are with respect to x. The Wightman function is a distribution on smooth functions, and thus when evaluating the integral in the definition of P_D the principle value prescription must be employed. If instead the field were in a different state, then the transition probability of the detector would still be given by Eq. (3.9) with the vacuum Wightman function replaced with the appropriate Wightman function describing the state of the field.

As discussed by Schlicht [38], we should expect a detector on a time-dependent trajectory or in a dynamic spacetime to react differently at different times. However, as constructed the transition probability has no such time dependence. For this reason, one may be interested in the transition rate of a detector which can depend on time. We define the transition rate of a detector as follows.

The probability that the detector has transitioned from its ground state to its excited state at the detector's proper time τ is given by Eq. (3.9) with the upper bounds of the integration replaced with τ

$$P_D(\tau) = \lambda^2 \int_{-\infty}^{\tau} dt \int_{-\infty}^{\tau} dt' \, \chi_D(t)\chi_D(t')e^{-i\Omega(t-t')} W\big(x_D(t), x_D(t')\big). \quad (3.12)$$

Following the approach of Schlicht [38], we introduce the integration variables $u := t$ and $s := t - t'$ for $t > t'$ and $u := t'$ and $s := t' - t$ for $t < t'$. This leads to

$$P_D(\tau) = \lambda^2 \int_{-\infty}^{\tau} du \int_{0}^{\infty} ds \, \chi_D(u)\chi_D(u - s)\Big[e^{-i\Omega s} W(x_D(u), x_D(u - s))$$

$$+ e^{i\Omega s} W(x_D(u - s), x_D(u))\Big]$$

$$= 2\lambda^2 \int_{-\infty}^{\tau} du \int_{0}^{\infty} ds \, \chi_D(u)\chi_D(u - s) \, \mathrm{Re}\Big[e^{-i\Omega s} W(x_D(u), x_D(u - s))\Big].$$

$$(3.13)$$

The transition rate $\dot{P}_D(\tau)$ is then defined as the derivative of the transition probability with respect to the proper time of the detector

$$\dot{P}_D(\tau) := \frac{d}{d\tau} P_D(\tau)$$

$$= 2\lambda^2 \int_{0}^{\infty} ds \, \chi_D(\tau)\chi_D(\tau - s) \, \mathrm{Re}\Big[e^{-i\Omega s} W(x_D(\tau), x_D(\tau - s))\Big].$$

$$(3.14)$$

In the limit where the switching function approaches the characteristic function on the interval $[\tau_0, \tau]$

$$\chi_D(t) = \begin{cases} 1 & \tau_0 \le t \le \tau \\ 0 & \text{otherwise} \end{cases}, \quad (3.15)$$

the transition rate becomes

$$\dot{P}_D(\tau) = \lambda^2 \left(\frac{1}{4} + 2 \int_0^{\Delta\tau} ds \ \text{Re}\left[e^{-i\Omega s} W(x_D(\tau), x_D(\tau - s)) \right] \right), \qquad (3.16)$$

where $\Delta\tau := \tau - \tau_0$. This limit was explicitly evaluated by Louko and Satz [24] by observing that when the Wightman function appearing under the integral is represented by an $i\epsilon$-regularized function, the regulator limit $\epsilon \to 0$ and the limit in which the switching function approaches the characteristic function (the sharp switching limit), do not in general commute and the first must be taken before the second.

3.2 Entanglement Harvesting in Curved Spacetimes

Entanglement is a uniquely quantum property of a composite system, such that the state of the entire system cannot be fully specified by the individual states of each component. Entanglement was first discussed by Einstein, Podolsky, and Rosen, who used an example of an entangled state to argue that quantum mechanics is incomplete [17]. Shortly after, Schrödinger introduced the term 'entanglement' [39], describing the phenomena as

> ...[not] *one*, but rather *the* characteristic trait of quantum mechanics, the one that enforces its entire departure from classical lines of thought.

In quantum computation, entanglement is used as a resource in many quantum protocols to give a dramatic speed-up over the corresponding classical protocol and can be used to construct unconditionally secure cryptographic systems [11, 29].

Formally, a quantum system composed of two subsystems A and B described by the Hilbert spaces $\mathcal{H}_A$ and $\mathcal{H}_B$, respectively, is associated with the Hilbert space $\mathcal{H}_A \otimes \mathcal{H}_B$. A state $\rho_{AB} \in \mathcal{S}(\mathcal{H}_A \otimes \mathcal{H}_B)$ is said to be factorized if it is of the form $\rho_{AB} = \rho_A \otimes \rho_B$. If ρ_{AB} is a convex combination of factorized states,

$$\rho_{AB} = \sum_i p_i \, \rho_A^{(i)} \otimes \rho_B^{(i)}, \qquad (3.17)$$

where $p_i > 0$ and $\sum_i p_i = 1$, then ρ_{AB} is said to be separable. If ρ_{AB} is not separable, it is entangled. These definitions of factorized, separable, and entangled states naturally generalize to systems composed of more than two subsystems [23]; however, such generalizations will not be needed for our purposes.

During the 1960s Bell, through the inequality that now bears his name [6, 7], placed an upper bound on the correlations predicted by any theory compatible with local realism.[3] He then demonstrated that the phenomenon of entanglement

---

[3] As stated in [1], a local realist theory is one where physical properties are defined prior to and independent of measurement, and no physical influence can propagate faster than the speed of light.

allows for the violation of his inequality, and consequently the model of local realism assumed by Einstein, Podolsky, and Rosen was incompatible with quantum mechanics. The first experimental violation of Bell's inequality was performed in 1972 by Freedman and Clauser [19], in which correlations between the polarization of photons emitted in an atomic cascade of calcium were measured and shown to violate Bell's inequality. Following this, numerous experiments have demonstrated the violation of Bell's inequality in a range of different experimental setups and have aimed at closing various loopholes [1].

Over the past three decades the role entanglement plays in quantum field theory has been explored, finding applications in disparate areas of physics including the study of critical phenomena in condensed matter systems [4, 30, 52], in the description of non-classical states of light within the field of quantum optics [25, 53], in explaining the origin of black hole entropy [10, 12, 43], and perhaps most spectacularly in the anti-de Sitter/conformal field theory correspondence in which the entanglement entropy associated with a region of a conformal field theory located on the boundary of anti-de Sitter space is related to minimal surfaces in the bulk [36].

Within the framework of algebraic quantum field theory, Summers and Werner [44–46] demonstrated that the vacuum state of a free quantum field in Minkowski space, as seen by local inertial observers, is entangled, and that the correlations seen by these observers are strong enough to violate Bell-type inequalities, even if these observers are in spacelike separated regions. This result is surprising—it suggests that no source of entanglement is necessary to detect a violation of Bell's inequality, the observation of vacuum fluctuations suffices. However, to the author's knowledge, algebraic quantum field theory has yet to address Bell inequalities in curved spacetimes [18, 21].

An operational approach to the study of the entanglement structure of the vacuum state of a quantum field theory was introduced in 1991 by Valentini [49]. He showed that two initially uncorrelated atoms separated by a distance R interacting with the electromagnetic vacuum exhibit nonlocal correlations after a time $t < R/c$. This implies that the atoms become entangled even if they remain spacelike separated throughout their interaction with the electromagnetic vacuum. He suggested that these correlations can either be interpreted as non-local photon propagation or as a consequence of non-locally correlated vacuum fluctuations.

In 2002 Reznik et al. [34, 35] demonstrated a similar effect using two Unruh-DeWitt detectors interacting locally with the vacuum state of a scalar field, and using three detectors showed that the Minkowski vacuum exhibits genuine non-local tripartite correlations, which are in principle strong enough to violate the Svetlichny inequality[4] [41].

This process of localized detectors extracting entanglement/nonlocal correlations from the vacuum state of a quantum field has since become known as entanglement

---

[4]The Svetlichny inequality is a Bell-like inequality whose violation is sufficient but not necessary for genuine tripartite nonlocal correlations [47].

harvesting [37], and has been studied in a variety of different situations, ranging from the extraction of resources from the vacuum [26], to entanglement generation between hydrogen-like atoms [32, 33], and even shown to depend on the underlying spacetime geometry [50] and topology [28]. We will refer to this process as the entanglement harvesting protocol.

The operational approach of Valentini and Reznik, in which the measurement of the field is explicitly described by an appropriate interaction with a measuring apparatus (atoms and Unruh-DeWitt detectors, respectively), emphasizes that the question of whether the vacuum state of a quantum field theory is seen to be entangled depends on the measurement process itself. This is analogous to how the notion of a particle depends on the measurement process, as discussed at the end of Chap. 2. In particular, whether the field is seen to be entangled can depend on the observer's motion. For example, Salton et al. [37] have demonstrated that the amount of entanglement in the Minkowski vacuum as seen by inertial detectors differs from that seen by uniformly accelerating detectors. This is analogous to the Unruh effect, in which uniformly accelerating observers disagree with inertial observers on the particle content of a given vacuum state.

The purpose of this section is to generalize the approach of Reznik et al. [34, 35] to Unruh-DeWitt detectors following arbitrary timelike trajectories through curved spacetimes admitting a Wightman function. This will allow for the study of the effect spacetime topology (Chap. 4) and gravity (Chap. 5) have on vacuum entanglement.

The advantage of probing the entanglement structure of a quantum field theory with Unruh-DeWitt detectors is (1) the approach is manifestly operational—the measuring process is explicitly described and the observer's motion can easily be taken into account; and (2) the mathematical machinery is simple as compared to algebraic quantum field theory and other methods [13, 14].

We consider two initially unexcited Unruh-DeWitt detectors which interact with a quantum field prepared in a suitably defined vacuum state for a finite period of time. The joint state of the two detectors after the interaction with the field has ceased is derived to all orders in the interaction strength between the detectors and field, and the leading order contribution is expressed in terms of the Wightman function. In general, this state can be entangled. Various measures of this entanglement are computed.

3.2.1 The Joint State of Two Detectors

To probe the entanglement structure of a scalar field, consider two Unruh-DeWitt detectors labelled by A and B described by the Hilbert spaces $\mathcal{H}_A$ and $\mathcal{H}_B$, respectively. Suppose these detectors follow the trajectories $x_A(\tau_A)$ and $x_B(\tau_B)$, which are parametrized by the detectors' proper times, τ_A and τ_B. These detectors couple to a real scalar field ϕ associated with the Hilbert space $\mathcal{H}_\phi$ via the interaction

Hamiltonians $H_A(\tau_A)$ and $H_B(\tau_B)$, which are given in Eq. (3.4). The evolution of these detectors and the field is described by the unitary operator

$$
U = \mathcal{T} \exp\left[-i \int dt \left(\frac{d\tau_A}{dt} H_A[\tau_A(t)] + \frac{d\tau_B}{dt} H_B[\tau_B(t)]\right)\right]
$$

$$
= 1 - i \int dt \left(\frac{d\tau_A}{dt} H_A[\tau_A(t)] + \frac{d\tau_B}{dt} H_B[\tau_B(t)]\right)
$$

$$
- \frac{1}{2} \int dt dt' \, \mathcal{T} \left[\frac{d\tau_A}{dt} \frac{d\tau_A}{dt'} H_A[\tau_A(t)] \, H_A[\tau_A(t')]\right.
$$

$$
+ \frac{d\tau_B}{dt} \frac{d\tau_B}{dt'} H_B[\tau_B(t)] \, H_B[\tau_B(t')]
$$

$$
+ \frac{d\tau_A}{dt} \frac{d\tau_B}{dt'} H_A[\tau_A(t)] \, H_B[\tau_B(t')]
$$

$$
+ \left. \frac{d\tau_B}{dt} \frac{d\tau_A}{dt'} H_B[\tau_B(t)] \, H_A[\tau_A(t')]\right] + \mathcal{O}\left(\lambda^3\right),
\tag{3.18}
$$

where we have chosen to evolve the field and detectors with respect to an appropriate coordinate time t with respect to which the vacuum state of the field is defined. The integrals appearing in Eq. (3.18) are over the interval $t \in (-\infty, \infty)$.

Suppose the two detectors are initially ($t \to -\infty$) prepared in their ground states, $|0\rangle_A$ and $|0\rangle_B$, and the field in an appropriately defined vacuum state $|0\rangle$, so that the initial state of the two detectors and field together is given by $|\Psi_i\rangle = |0\rangle_A |0\rangle_B |0\rangle \in \mathcal{H}_A \otimes \mathcal{H}_B \otimes \mathcal{H}_\phi$. After the interaction ($t \to \infty$), the joint state of the two detectors and field is

$$
|\Psi_f\rangle = U |\Psi_i\rangle = \sum_n \lambda^n \left|\Psi_f^{(n)}\right\rangle,
\tag{3.19}
$$

where $\left|\Psi_f^{(n)}\right\rangle$ is the nth order contribution to the final state $|\Psi_f\rangle \in \mathcal{H}_A \otimes \mathcal{H}_B \otimes \mathcal{H}_\phi$. Defining $\eta_D(t) := \chi_D(\tau_D(t)) d\tau_D/dt$ and $\phi_D(t) := \phi[x_D(t)]$ where $D \in \{A, B\}$, the first three terms in Eq. (3.19) are given below.

Zeroth-Order in λ

$$
\left|\Psi_f^{(0)}\right\rangle = |0\rangle_A |0\rangle_B |0\rangle .
\tag{3.20}
$$

First-Order in λ

$$
\left|\Psi_f^{(1)}\right\rangle = -i \int dt \left(\eta_A(t) e^{i\Omega_A \tau_A(t)} |1\rangle_A |0\rangle_B \otimes \phi_A(t) |0\rangle\right.
$$

$$
\left. + \eta_B(t) e^{i\Omega_B \tau_B(t)} |0\rangle_A |1\rangle_B \otimes \phi_B(t) |0\rangle\right).
\tag{3.21}
$$

Second-Order in λ

$$\left|\Psi_f^{(2)}\right\rangle = -\frac{1}{2}\int dt\,dt'$$

$$\left(\eta_A(t)\eta_A(t')\mathcal{T}\left[\sigma_A^-(t)\sigma_A^+(t') + \sigma_A^+(t)\sigma_A^-(t')\right]\left|0\right\rangle_A\left|0\right\rangle_B\right.$$

$$\otimes\,\mathcal{T}\phi_A(t)\phi_A(t')\left|0\right\rangle$$

$$+\,\eta_B(t)\eta_B(t')\left|0\right\rangle_A\otimes\mathcal{T}\left[\sigma_B^-(t)\sigma_B^+(t') + \sigma_B^+(t)\sigma_B^-(t')\right]\left|0\right\rangle_B$$

$$\left.\otimes\,\mathcal{T}\phi_A(t)\phi_A(t')\left|0\right\rangle\right)$$

$$-\frac{1}{2}\int dt\,dt'\,\eta_A(t)\eta_B(t')e^{i\left[\Omega_A\tau_A(t)+\Omega_B\tau_B(t')\right]}\left|1\right\rangle_A\left|1\right\rangle_B$$

$$\otimes\left(\mathcal{T}\phi_B(t')\phi_A(t) + \mathcal{T}\phi_A(t)\phi_B(t')\right)\left|0\right\rangle. \tag{3.22}$$

The reduced state of the two detectors $\rho_{AB} \in \mathcal{S}\left(\mathcal{H}_A\otimes\mathcal{H}_B\right)$ after their interaction with the field has ceased is obtained from $\left|\Psi_f\right\rangle$ by tracing out the field

$$\rho_{AB} := \mathrm{tr}_\phi\left(\left|\Psi_f\right\rangle\!\left\langle\Psi_f\right|\right) = \sum_{n,m}\lambda^{n+m}\int d\mu\,\langle\mu|\left(\left|\Psi_f^{(n)}\right\rangle\!\left\langle\Psi_f^{(m)}\right|\right)|\mu\rangle, \tag{3.23}$$

where $|\mu\rangle$ is an element of the Fock basis of the field Hilbert space, which is used to perform the partial trace.

Observe that for even (odd) n, the state $\left|\Psi^{(n)}(t_f)\right\rangle$ has an even (odd) number of excitations in the field since the field operator $\phi_D(\cdot)$ is applied to the field vacuum $|0\rangle$ an even (odd) number of times. As field states with an even number of excitations are orthogonal to field states with an odd number of excitations, only terms where both n and m are either even or odd survive the partial trace in Eq. (3.23). Also observe that for even n, either both detectors are excited or unexcited in the state $\left|\Psi^{(n)}(t_f)\right\rangle$; for odd n, one detector is excited and the other is not in the state $\left|\Psi^{(n)}(t_f)\right\rangle$. Combining these observations [28], one concludes that the reduced density matrix of the two detectors to all orders of perturbation theory, in the basis $\{\left|0\right\rangle_A\left|0\right\rangle_B, \left|0\right\rangle_A\left|1\right\rangle_B, \left|1\right\rangle_A\left|0\right\rangle_B, \left|1\right\rangle_A\left|1\right\rangle_B\}$, is

$$\rho_{AB} = \begin{pmatrix} \rho_{11} & 0 & 0 & \rho_{14} \\ 0 & \rho_{22} & \rho_{23} & 0 \\ 0 & \rho_{23}^* & \rho_{33} & 0 \\ \rho_{14}^* & 0 & 0 & \rho_{44} \end{pmatrix} = \begin{pmatrix} 1-P_A-P_B & 0 & 0 & X \\ 0 & P_B & C & 0 \\ 0 & C^* & P_A & 0 \\ X^* & 0 & 0 & 0 \end{pmatrix} + \mathcal{O}\left(\lambda^4\right), \tag{3.24}$$

where P_A and P_B are the transition probabilities of detectors A and B defined in Eq. (3.9), and

$$C := \lambda^2 \int dt dt' \, \eta_B(t) \eta_A(t') e^{i[\Omega_B \tau_B(t) - \Omega_A \tau_A(t')]} W\big(x_A(t'), x_B(t)\big), \qquad (3.25)$$

$$X := -\lambda^2 \int_{t>t'} dt dt' \bigg[\eta_B(t) \eta_A(t') e^{-i[\Omega_B \tau_B(t) + \Omega_A \tau_A(t')]} W\big(x_A(t'), x_B(t)\big)$$

$$+ \eta_A(t) \eta_B(t') e^{-i[\Omega_A \tau_A(t) + \Omega_B \tau_B(t')]} W\big(x_B(t'), x_A(t)\big) \bigg]. \qquad (3.26)$$

The leading order contribution to $\rho_{44} = E + \mathcal{O}(\lambda^6)$ is denoted by E, which is derived in detail in Appendix A with the result

$$E = |X|^2 + |C|^2 + P_A P_B. \qquad (3.27)$$

The final state ρ_{AB} of the detectors is an example of an X-state [3], the name being due to the density matrix's resemblance with the letter 'X'.

The reduced state of detector A is

$$\rho_A := \mathrm{tr}_B \, \rho_{AB} = \begin{pmatrix} \rho_{11} + \rho_{22} & 0 \\ 0 & \rho_{33} + \rho_{44} \end{pmatrix} = \begin{pmatrix} 1 - P_A & 0 \\ 0 & P_A \end{pmatrix} + \mathcal{O}(\lambda^4), \qquad (3.28)$$

and the reduced state detector B is

$$\rho_B := \mathrm{tr}_A \, \rho_{AB} = \begin{pmatrix} \rho_{11} + \rho_{33} & 0 \\ 0 & \rho_{22} + \rho_{44} \end{pmatrix} = \begin{pmatrix} 1 - P_B & 0 \\ 0 & P_B \end{pmatrix} + \mathcal{O}(\lambda^4). \qquad (3.29)$$

The reduced states $\rho_A \in \mathcal{S}(\mathcal{H}_A)$ and $\rho_B \in \mathcal{S}(\mathcal{H}_B)$ coincide with the density matrix of the single detector in Eq. (3.8). This is an important consistency requirement—if this was not the case, it would imply that measurements of either detector could infer that another potentially spacelike separated detector is interacting with the field, allowing for the possibility of superluminal signalling.

3.2.2   Quantifying the Entanglement in ρ_{AB}

Having derived the state $\rho_{AB} \in \mathcal{S}(\mathcal{H}_A \otimes \mathcal{H}_B)$ of two Unruh-DeWitt detectors after their interaction with the vacuum state of a scalar field, Eq. (3.24), we now wish to quantify how entangled this state is. If the detectors remain spacelike separated throughout their entire interaction with the field, then any entanglement that results between the detectors must have been extracted from the vacuum since the detectors would not have been able to interact directly. In this situation, the entanglement present in ρ_{AB} is an indicator of the entanglement between the regions of the field with which the detectors interacted.

If the detectors do not remain spacelike separated for the duration of their interaction with the field, the entanglement present in ρ_{AB} may still be a result of entanglement extracted from the vacuum, but it may also be a result of an interaction between the detectors mediated by the field. Consequently, in this case one cannot conclude that the entanglement contained in ρ_{AB} has been extracted from the vacuum alone. However, as we will see in Chaps. 4 and 5 the entanglement that results when the detectors are not spacelike separated can still depend on the properties of the spacetime itself.

As a first observation, note that C and X defined in Eqs. (3.25) and (3.26) are an integration over the Wightman function $W(x, x')$ evaluated along the trajectories of the detectors $x_A(\tau_A)$ and $x_B(\tau_B)$. If $W(x, x')$ vanishes for spacetime points x and x' infinitely far apart, as is the case in Minkowski space, then for detectors infinitely far apart C and X vanish too, and the final state of the two detectors factorizes

$$\rho_{AB} = \rho_A \otimes \rho_B. \tag{3.30}$$

Quantifying the amount of entanglement in a given state is done by the use of entanglement measures [11]. Given a state $\rho_{AB} \in \mathcal{S}(\mathcal{H}_A \otimes \mathcal{H}_B)$, an entanglement measure $E(\rho_{AB})$ is a map

$$E : \mathcal{S}(\mathcal{H}_A \otimes \mathcal{H}_B) \rightarrow \mathbb{R}^+, \tag{3.31}$$

such that $E(\rho_{AB}) = 0$ if ρ_{AB} is separable, and $E(\rho_{AB})$ does not increase on average under local operations on $\mathcal{H}_A$ and $\mathcal{H}_B$ and classical communication between the parties associated with $\mathcal{H}_A$ and $\mathcal{H}_B$.

In general, the final state of the two detectors ρ_{AB} will be mixed, and therefore the entanglement entropy is not a suitable measure of entanglement [11]. Consequently, in what follows we introduce three entanglement measures that are suitable to characterize the entanglement in the state ρ_{AB}, as well as a measure of correlations in this state, and explicitly evaluate them for the density matrix given in Eq. (3.24).

Negativity

The Peres-Horodecki criterion asserts that a state $\rho_{AB} \in \mathcal{S}(\mathcal{H}_A \otimes \mathcal{H}_B)$ is entangled if ρ_{AB} does not remain positive under partial transposition[5] with respect to either $\mathcal{H}_A$ or $\mathcal{H}_B$. If $\mathcal{H}_A \simeq \mathbb{C}^2$ and $\mathcal{H}_B \in \{\mathbb{C}^2, \mathbb{C}^3\}$, then ρ_{AB} is entangled if and only if it does not remain positive under partial transposition with respect to either $\mathcal{H}_A$ or $\mathcal{H}_B$ [22, 31].

The negativity is defined as [51]

$$\mathcal{N}(\rho_{AB}) := \frac{\left\| \rho_{AB}^{\Gamma_A} \right\| - 1}{2} = \sum_{\lambda_i < 0} |\lambda_i|, \tag{3.32}$$

---

[5]The partial transpose of the state $\rho_{AB} \in \mathcal{S}(\mathcal{H}_A \otimes \mathcal{H}_B)$ with respect to $\mathcal{H}_A$ is $\rho_{AB}^{\Gamma_A} :=$ $[T \otimes \mathcal{I}](\rho_{AB})$, where $T : \mathcal{S}(\mathcal{H}_A) \rightarrow \mathcal{S}(\mathcal{H}_A)$ is the transposition map.

where Γ_A denotes the partial transpose with respect to A, $\|\cdot\|$ denotes the trace norm, and the sum in the last equality is over the negative eigenvalues λ_i of $\rho_{AB}^{\Gamma_A}$. The negativity quantifies the degree to which $\rho_{AB}^{\Gamma_A}$ fails to be positive [11, 29]. The negativity vanishes on separable states and does not increase under local operations and classical communication, and therefore the negativity is an entanglement measure.

The partial transpose of the two-detector density matrix given in Eq. (3.24) is

$$
\rho_{AB}^{\Gamma_A} = \begin{pmatrix} \rho_{11} & 0 & 0 & \rho_{23}^* \\ 0 & \rho_{22} & \rho_{14}^* & 0 \\ 0 & \rho_{14} & \rho_{33} & 0 \\ \rho_{23} & 0 & 0 & \rho_{44} \end{pmatrix},
\tag{3.33}
$$

with eigenvalues

$$
\begin{aligned}
\lambda_1 &= \frac{\rho_{22} + \rho_{33}}{2} - \sqrt{\left(\frac{\rho_{22} - \rho_{33}}{2}\right)^2 + |\rho_{14}|^2} \\
&= \frac{P_A + P_B}{2} - \sqrt{\left(\frac{P_A - P_B}{2}\right)^2 + |X|^2} + \mathcal{O}\left(\lambda^4\right),
\end{aligned}
\tag{3.34a}
$$

$$
\begin{aligned}
\lambda_2 &= \frac{\rho_{22} + \rho_{33}}{2} + \sqrt{\left(\frac{\rho_{22} - \rho_{33}}{2}\right)^2 + |\rho_{14}|^2} \\
&= \frac{P_A + P_B}{2} + \sqrt{\left(\frac{P_A - P_B}{2}\right)^2 + |X|^2} + \mathcal{O}\left(\lambda^4\right),
\end{aligned}
\tag{3.34b}
$$

$$
\begin{aligned}
\lambda_3 &= \frac{\rho_{11} + \rho_{44}}{2} - \sqrt{\left(\frac{\rho_{11} - \rho_{44}}{2}\right)^2 + |\rho_{23}|^2} \\
&= |X|^2 + P_A P_B + \mathcal{O}\left(\lambda^6\right),
\end{aligned}
\tag{3.34c}
$$

$$
\begin{aligned}
\lambda_4 &= \frac{\rho_{11} + \rho_{44}}{2} + \sqrt{\left(\frac{\rho_{11} - \rho_{44}}{2}\right)^2 + |\rho_{23}|^2} \\
&= 1 - P_A - P_B + \mathcal{O}\left(\lambda^4\right).
\end{aligned}
\tag{3.34d}
$$

The only possible negative eigenvalue of $\rho_{AB}^{\Gamma_A}$ is λ_1, and thus, by the Peres-Horodecki criterion, ρ_{AB} is entangled if and only if

$$
\lambda_1 < 0 \quad \Rightarrow \quad |\rho_{14}|^2 > \rho_{22}\rho_{33} \quad \Rightarrow \quad |X|^2 > P_A P_B.
\tag{3.35}
$$

Note that the second equality in Eq. (3.34c), and consequently the above condition, required knowledge of the leading order contribution to ρ_{44} given in Eq. (3.27), which is $\mathcal{O}(\lambda^4)$ [28].

Applying Eq. (3.36), the negativity of the final state of the two detectors in Eq. (3.24) is

$$
\mathcal{N}(\rho_{AB}) = \max\left[0, \sqrt{|X|^2 + \left(\frac{P_A - P_B}{2}\right)^2} - \frac{P_A + P_B}{2}\right] + \mathcal{O}(\lambda^4). \quad (3.36)
$$

The advantage of using the negativity to quantify entanglement over other measures is that it is comparatively easy to compute. Consequently, the literature on entanglement harvesting has focused exclusively (with the exception of [28]) on the negativity to quantify the entanglement that results between two Unruh-DeWitt detectors. However, the negativity does not have a direct operational interpretation. For this reason, we consider the entanglement of formation in the next section.

Concurrence and the Entanglement of Formation

The entanglement of formation $E_f(\rho_{AB})$ is an entanglement measure for bipartite quantum states $\rho_{AB} \in \mathcal{S}(\mathcal{H}_A \otimes \mathcal{H}_B)$ defined as the lowest entanglement entropy of any ensemble realizing ρ_{AB} [8], explicitly

$$
E_f(\rho_{AB}) := \min \sum_i p_i E(\psi_i), \quad (3.37)
$$

where the minimization is carried out over all pure state decompositions of ρ_{AB}, that is, the ensembles $\{p_i, |\psi_i\rangle\}$ such that $\rho_{AB} := \sum_i p_i |\psi_i\rangle\langle\psi_i|$; $E(\psi_i)$ is the von Neumann entropy of either of the two subsystems

$$
E(\psi_i) := -\operatorname{tr} \rho_{A_i} \log \rho_{A_i} = -\operatorname{tr} \rho_{B_i} \log \rho_{B_i}, \quad (3.38)
$$

where $\rho_{A_i} := \operatorname{tr}_B |\psi_i\rangle\langle\psi_i|$ and $\rho_{B_i} := \operatorname{tr}_A |\psi_i\rangle\langle\psi_i|$. The entanglement of formation has an operational interpretation as the number of Bell states required to prepare ρ_{AB} via local operations and classical communication [8].

The solution to the minimization problem defining the entanglement of formation is [54]

$$
E_f(\rho_{AB}) := h\left(\frac{1 + \sqrt{1 - \mathcal{C}(\rho_{AB})^2}}{2}\right), \quad (3.39)
$$

where $h(x) := -x \log x - (1 - x) \log(1 - x)$, and $\mathcal{C}(\rho_{AB})$ is the concurrence

$$
\mathcal{C}(\rho_{AB}) := \max[0, \, w_1 - w_2 - w_3 - w_4], \quad (3.40)
$$

where the w_i's are the square roots of the eigenvalues of the matrix $\rho_{AB}\tilde{\rho}_{AB}$; $\tilde{\rho}_{AB} := (\sigma_y \otimes \sigma_y) \rho_{AB}^* (\sigma_y \otimes \sigma_y)$ and σ_y is the Pauli y matrix. As the entanglement of formation is a monotonically increasing function of the concurrence and ranges from 0 to 1 as the concurrence goes from 0 to 1, the concurrence too is a measure of entanglement [54].

For the joint state ρ_{AB} of the two detectors given in Eq. (3.24), the square root of the eigenvalues of $\rho_{AB}\tilde{\rho}_{AB}$ is

$$w_1 = \sqrt{\rho_{11}\rho_{44}} + |\rho_{14}| = \sqrt{|X|^2 + |C|^2 + P_A P_B} + |X| + \mathcal{O}\left(\lambda^4\right), \qquad (3.41a)$$

$$w_2 = \sqrt{\rho_{22}\rho_{33}} + |\rho_{23}| = \sqrt{P_A P_B} + |C| + \mathcal{O}\left(\lambda^4\right), \qquad (3.41b)$$

$$w_3 = \sqrt{\rho_{11}\rho_{44}} - |\rho_{14}| = \sqrt{|X|^2 + |C|^2 + P_A P_B} - |X| + \mathcal{O}\left(\lambda^4\right), \qquad (3.41c)$$

$$w_4 = \sqrt{\rho_{22}\rho_{33}} - |\rho_{23}| = \sqrt{P_A P_B} - |C| + \mathcal{O}\left(\lambda^4\right). \qquad (3.41d)$$

If ρ_{AB} is entangled, the Peres-Horodecki criterion, $|X|^2 > P_A P_B$, implies that the largest eigenvalue of $\rho_{AB}\tilde{\rho}_{AB}$ is w_1. Using Eq. (3.40), the concurrence is seen to be

$$\mathcal{C}(\rho_{AB}) = 2 \max\left[0, \ |\rho_{14}| - \sqrt{\rho_{22}\rho_{33}}\right] = 2 \max\left[0, \ |X| - \sqrt{P_A P_B}\right] + \mathcal{O}\left(\lambda^4\right). \qquad (3.42)$$

If the transition probabilities of the two detectors are equal, $P_A = P_B$, the concurrence of ρ_{AB} is given by twice the negativity, $\mathcal{C}(\rho_{AB}) = 2\mathcal{N}(\rho_{AB})$ [28].

From the expression for the concurrence given in Eq. (3.42), we see that the final state of the two detectors ρ_{AB} is most entangled when the product of the transition probabilities of both detectors is small and the absolute value of the off diagonal element X is large. This observation will be important for the interpretation of the results presented in Chaps. 4 and 5.

Correlations

While the negativity, concurrence, and entanglement of formation are useful for quantifying entanglement, they are not directly accessible by local measurements of the detectors. Consequently, one may be interested in the correlation between the outcomes of local measurements of each detector.

As the final state of either detector is diagonal in the $\{|0\rangle_D, |1\rangle_D\}$ basis, Eqs. (3.28) and (3.29), the only nontrivial measurement is in this basis. The correlation between outcomes of these measurements is defined as

$$\text{corr}\,(\rho_{AB}) := \frac{\text{cov}\,(\rho_{AB})}{\text{var}\,(\rho_A)\,\text{var}\,(\rho_B)}, \qquad (3.43)$$

where

$$\mathrm{cov}\,(\rho_{AB}) := \mathrm{tr}\,(\rho_{AB}\,\sigma_z \otimes \sigma_z) - \mathrm{tr}\,(\rho_A \sigma_z)\,\mathrm{tr}\,(\rho_B \sigma_z)\,,$$

$$\mathrm{var}\,(\rho_D) := \mathrm{tr}\left(\rho_D \sigma_z^2\right) - \mathrm{tr}\,(\rho_D \sigma_z)^2\,,$$

for $D \in \{A, B\}$; $\mathrm{corr}\,(\rho_{AB})$ will be referred to as the correlation function.

Evaluating $\mathrm{corr}\,(\rho_{AB})$ for the final state of the two detectors ρ_{AB} in Eq. (3.24) yields

$$\begin{aligned}
\mathrm{corr}\,(\rho_{AB}) &= \frac{\rho_{44} - (\rho_{22} + \rho_{44})\,(\rho_{33} + \rho_{44})}{\sqrt{(1 - \rho_{22} - \rho_{44})\,(\rho_{22} + \rho_{44})\,(1 - \rho_{33} - \rho_{44})\,(\rho_{33} + \rho_{44})}} \\
&= \frac{|X|^2 + |C|^2}{\sqrt{P_A P_B}} + \mathcal{O}\!\left(\lambda^4\right).
\end{aligned} \tag{3.44}$$

3.3   Detector Observables

Sections 3.1 and 3.2 have emphasized that an Unruh-DeWitt detector, through the interaction Hamiltonian in Eq. (3.4), performs a measurement of a quantum field. The purpose of this section is to identify the measurement model a collection of Unruh-DeWitt detectors defines, and to identify the field observables[6] these detectors measure as a function of their initial state and trajectories.

A measurement procedure begins by coupling the system that is to be measured with a probe system. After some time, the system and probe are decoupled, and a measurement is carried out on the probe system only. As a result of the coupling stage, the system and probe become correlated, and a measurement of the probe system gives information about the system that is to be measured.

Following [20], we will now formalize this measurement procedure. Let A be an observable we wish to measure on a system associated with the Hilbert space $\mathcal{H}$, with possible outcomes $A(X) \in \mathcal{E}(\mathcal{H})$ where X labels the possible outcomes of A; $A(X)$ will be referred to as the POVM element associated with the outcome X. The collection of POVM elements $\{A(X)\ \forall X\}$ defines the observable A.

Suppose that we prepare a probe system in the state $|\xi\rangle \in \mathcal{K}$, where $\mathcal{K}$ is the Hilbert space associated with the probe system, and the interaction between the probe and system to be measured is described by a completely positive trace

---

[6]An observable is defined on a measurable space $(\Omega, \mathcal{F})$, where Ω is a sample space and $\mathcal{F}$ is a collection of subsets of Ω ($\mathcal{F}$ is a σ-algebra); $(\Omega, \mathcal{F})$ is the outcome space of the observable. An observable A is a map $A : \mathcal{F} \to \mathcal{E}(\mathcal{H})$, where $\mathcal{F}$ is the space of possible outcomes of a measurement of the observable A on a system associated with the Hilbert space $\mathcal{H}$, such that A is a positive operator valued measure (POVM). A POVM is a map $A : \mathcal{F} \to \mathcal{E}(\mathcal{H})$ such that (a) $A(\emptyset) = 0$, (b) $A(\Omega) = I$, and (c) $A(\cup_i X_i) = \sum_i A(X_i)$ for any sequence $\{X_i\}$ of disjoint sets in $\mathcal{F}$; for $X \in \mathcal{F}$, $A(X)$ are referred to as POVM elements. In other words, a mapping $A : \mathcal{F} \to \mathcal{E}(\mathcal{H})$ is a POVM if and only if the mapping $X \mapsto \mathrm{tr}\,[\rho A(X)]$ defines a probability measure for every state $\rho \in \mathcal{S}(\mathcal{H})$.

preserving map

$$V : S(\mathcal{K} \otimes \mathcal{H}) \rightarrow S(\mathcal{K} \otimes \mathcal{H}). \tag{3.45}$$

Suppose that after the interaction has ceased, we measure the observable F with outcomes $F(X) \in \mathcal{E}(\mathcal{K})$ on the probe system. The quadruple $\mathcal{M} = \{\mathcal{K}, |\xi\rangle, \mathcal{V}, F\}$ defines a measurement model of the observable A if it satisfies the probability reproducibility condition [20]

$$\mathrm{tr}\left[\rho A(X)\right] = \mathrm{tr}\left[\mathcal{V}\left[\,|\xi\rangle\langle\xi| \otimes \rho\right](F(X) \otimes I)\right], \tag{3.46}$$

for all $F(X)$ and $\rho \in S(\mathcal{H})$. This condition asserts that a measurement F of the probe, after the interaction with the system to be measured has ceased, results in the same probabilities as if a measurement of A was performed directly on the system.

However, it is important to note that the probability reproducibility condition can be applied in the opposite direction: a given measurement model $\mathcal{M} = \{\mathcal{K}, |\xi\rangle, \mathcal{V}, F\}$ defines an observable A on $\mathcal{H}$ given by

$$A(X) := \langle\xi|\mathcal{V}^*\left[F(X) \otimes I\right]|\xi\rangle \in \mathcal{E}(\mathcal{H}), \tag{3.47}$$

where $\mathcal{V}^*$ is the dual of $\mathcal{V}$. We will apply the probability reproducibility condition in this manner to answer the question: Which field observables can a collection of Unruh-DeWitt detectors measure?

Consider a collection of N detectors associated with the Hilbert space $\mathcal{K} = \bigotimes_i \mathcal{H}_i$, where $\mathcal{H}_i$ is the Hilbert space describing the ith detector. Suppose these detectors are used collectively as a probe system to make a measurement of a scalar field ϕ associated with the Hilbert space $\mathcal{H}_\phi$. Suppose that prior to the interaction with the field the detectors are prepared in the state $|\xi\rangle \in \mathcal{K}$ and the interaction between the probe and field is described by the interaction Hamiltonian

$$H_I(t) = \lambda \sum_i \mu_i(t) \otimes g[x_i(t)], \tag{3.48}$$

where $\mu_i(t) \in \mathcal{L}_s(\mathcal{H}_i)$ is a self-adjoint operator on $\mathcal{H}_i$, $g[x_i(t)] = g[\phi[x_i(t)], \pi[x_i(t)]] \in \mathcal{L}_s(\mathcal{H})$ is a function of the field operator ϕ and its conjugate momentum π evaluated along the ith detector's trajectory $x_i(t)$, and t is a suitably chosen coordinate time. For the Unruh-DeWitt detector considered in Sects. 3.1 and 3.2, $\mathcal{H}_i \simeq \mathbb{C}^2$, $\mu_i(t) = \eta_i(t)\left[\sigma^+(t) + \sigma^-(t)\right]$, and $g[x_i(t)] = \phi[x_i(t)]$. The channel describing the measurement interaction is $\mathcal{V}[\,\cdot\,] := U[\,\cdot\,]U^\dagger$, where

$$U = \mathcal{T}\exp\left[-i\lambda \int dt \sum_i \mu_i(t) \otimes g[x_i(t)]\right]$$

$$= I + (-i)\lambda \sum_i \int dt\, \mu_i(t) \otimes g[x_i(t)]$$

$$+ \frac{(-i)^2}{2}\lambda^2 \sum_{i,j} \int dt dt'\, \mathcal{T}\left(\mu_i(t)\mu_j(t') \otimes g[x_i(t)]g[x_j(t')]\right) + \mathcal{O}(\lambda^3).$$

$$\tag{3.49}$$

Finally, after the interaction with the field has ceased, suppose a measurement of the observable F, with outcomes $F(X) \in \mathcal{E}(\mathcal{K})$, is made on the final state of the detectors. The set

$$\mathcal{M} = \left\{ \mathcal{K} = \bigotimes_i \mathcal{H}_i, \ |\xi\rangle \in \mathcal{K}, \ \mathcal{V}[\cdot] = U[\cdot]U^\dagger, \ F \right\}, \tag{3.50}$$

constitutes a measurement model and in turn defines an observable A on the scalar field ϕ with outcomes $A(X) \in \mathcal{E}(\mathcal{H})$ given by Eq. (3.47)

$$A(X) := \left\langle \xi \middle| U^\dagger \big(F(X) \otimes I\big) U \middle| \xi \right\rangle$$

$$= \langle \xi | F(X) | \xi \rangle \, I + i\lambda \int dt \sum_i \langle \xi | \, [\mu_i(t) \otimes g[x_i(t)], F(X) \otimes I] \, | \xi \rangle$$

$$+ \frac{(i\lambda)^2}{2!} \int dt\,dt'$$

$$\times \sum_{j,i} \mathcal{T} \big\langle \xi | \, [\mu_i(t) \otimes g[x_i(t)], [\mu_j(t) \otimes g[x_j(t)], F(X) \otimes I]] \, | \xi \big\rangle$$

$$+ \mathcal{O}\big(\lambda^3\big)$$

$$= \langle \xi | F(X) | \xi \rangle \, I + i\lambda \int dt \sum_i \langle \xi | \, [\mu_i(t), F(X)] \, | \xi \rangle \, g[x_i(t)]$$

$$+ \frac{(i\lambda)^2}{2!} \int dt'\,dt$$

$$\times \sum_{j,i} \mathcal{T} \big\langle \xi | \, [\mu_j(t') \otimes g[x_j(t')], [\mu_i(t), F(X)] \otimes g[x_i(t)]] \, | \xi \big\rangle + \mathcal{O}\big(\lambda^3\big)$$

$$= \langle \xi | F(X) | \xi \rangle \, I + i\lambda \int dt \sum_i \langle \xi | \, [\mu_i(t), F(X)] \, | \xi \rangle \, g[x_i(t)]$$

$$+ \frac{(i\lambda)^2}{2!} \int dt'\,dt \sum_{j,i} \mathcal{T} \Big(\langle \xi | \mu_j(t') \, [\mu_i(t), F(X)] \, | \xi \rangle \, g[x_j(t')]g[x_i(t)]$$

$$- \langle \xi | \, [\mu_i(t), F(X)] \, \mu_j(t') | \xi \rangle \, g[x_i(t)]g[x_j(t')] \Big) + \mathcal{O}\big(\lambda^3\big). \tag{3.51}$$

Swapping the integration variables, $t \leftrightarrow t'$, and summation indices, $i \leftrightarrow j$, in the second order term and assuming $[\mu_i(t), \mu_j(t')] = 0$, as is the case for Unruh-DeWitt detectors, the second order term in Eq. (3.51) simplifies to

$$(i\lambda)^2 \int dt'\,dt \sum_{i,j} \mathrm{Re}\left[\langle \xi | \mu_j(t') \, [\mu_i(t), F(X)] \, | \xi \rangle \right] \mathcal{T} g[x_j(t')]g[x_i(t)]. \tag{3.52}$$

Defining the functions

$$f^{(0)}(X) := \langle \xi | F(X) | \xi \rangle, \tag{3.53a}$$

$$f_i^{(1)}(X;t) := i \langle \xi | [\mu_i(t), F(X)] | \xi \rangle, \tag{3.53b}$$

$$f_{ji}^{(2)}(X;t',t) := i^2 \operatorname{Re} \left[\langle \xi | \mu_j(t') [\mu_i(t), F(X)] | \xi \rangle \right], \tag{3.53c}$$

$A(X)$ becomes

$$A(X) = f^{(0)}(X)I + \lambda \sum_i \int dt \, f_i^{(1)}(X;t) g[x_i(t)]$$

$$+ \lambda^2 \sum_{j,i} \int dt' dt \, f_{ji}^{(2)}(X;t',t) \mathcal{T} g[x_j(t')] g[x_i(t)] + \mathcal{O}\left(\lambda^3\right). \tag{3.54}$$

From Eq. (3.54), we see that at zeroth order in the interaction strength $A(X)$ is proportional to the identity; the first order contribution to $A(X)$ is the operator $g[x_i(t)]$ appearing in the interaction Hamiltonian in Eq. (3.48) smeared over the detectors' trajectories by the function $f_i^{(1)}(X;t)$; and the second order contribution is the time ordered product $\mathcal{T} g[x_j(t')] g[x_i(t)]$ smeared over the detectors trajectories with the function $f_{ji}^{(2)}(X;t',t)$. Note that the analysis above was not specific to Unruh-DeWitt detectors.

Having now constructed the field observable A defined by a collection of detectors prepared in an arbitrary pure state $|\xi\rangle$ moving along trajectories $x_i(t)$, through the probability reproducibility condition, Eq. (3.46), one can easily compute the probability of different outcomes X of various measurements of observables F on the collection of detectors.

3.3.1 Example: Single Unruh-DeWitt Detector Observables

As an example of an application of the above measurement model, let us compute the observables measured by a single Unruh-DeWitt detector moving along the trajectory $x_D(t)$. As was done in Sect. 3.1, let us consider the detector to be initially in its ground state $|\xi\rangle = |0\rangle_D \in \mathcal{K} \simeq \mathbb{C}^2$. The interaction between the scalar field and detector will be given by Eq. (3.48), which corresponds to $g[x_D(t)] = \phi[x_D(t)]$ and

$$\mu_D(t) := \eta_D(t) \left(e^{i\Omega \tau(t)} \sigma^+ + e^{-i\Omega \tau(t)} \sigma^- \right), \tag{3.55}$$

where $\sigma^+ = |1\rangle_D \langle 0|_D$ and $\sigma^- = |0\rangle_D \langle 1|_D$. Suppose that after the interaction with the field, the measurement of the detector is a projective measurement described by the POVM elements

$$F := \left\{ P_1 := |a\rangle_D\langle a|_D \,, \ P_2 := I - |a\rangle_D\langle a|_D \right\} \in \mathcal{E}(\mathcal{K}), \tag{3.56}$$

for some $|a\rangle \in \mathcal{K}$, with outcomes X_1 and X_2 corresponding to P_1 and P_2, respectively. As the measurement of the detector has two outcomes, so will the field observable $A(X)$ defined by this measurement model.

To compute the POVM elements $A(X)$ defining A, we must evaluate the smearing functions given in Eq. (3.53). Beginning with $f^{(0)}(X)$

$$f^{(0)}(X_1) := \langle \xi | F(X_1) | \xi \rangle = \langle 0 | P_1 | 0 \rangle = |\langle a | 0 \rangle|^2 \,, \tag{3.57a}$$

$$f^{(0)}(X_2) := \langle \xi | F(X_2) | \xi \rangle = \langle 0 | P_2 | 0 \rangle = 1 - |\langle a | 0 \rangle|^2 \,, \tag{3.57b}$$

where we have dropped the subscript D for clarity. The first order smearing function $f^{(1)}(X; t)$ is

$$
\begin{aligned}
f^{(1)}(X_1; t) &:= i \langle \xi | [\mu_D(t), F(X_1)] | \xi \rangle \\
&= i\eta_D(t)\left(e^{i\Omega\tau(t)} \langle 0 | [\sigma^+, P_1] | 0 \rangle + e^{-i\Omega\tau(t)} \langle 0 | [\sigma^-, P_1] | 0 \rangle \right) \\
&= i\eta_D(t)\left(-e^{i\Omega\tau(t)} \langle 0 | a \rangle \langle a | 1 \rangle + e^{-i\Omega\tau(t)} \langle 1 | a \rangle \langle a | 0 \rangle \right) \\
&= 2\eta_D(t) \operatorname{Im}\left[e^{i\Omega\tau(t)} \langle 0 | a \rangle \langle a | 1 \rangle \right],
\end{aligned}
\tag{3.58a}
$$

$$
\begin{aligned}
f^{(1)}(X_2; t) &:= i \langle \xi | [\mu_D(t), F(X_2)] | \xi \rangle \\
&= -2\eta_D(t) \operatorname{Im}\left[e^{i\Omega\tau(t)} \langle 0 | a \rangle \langle a | 1 \rangle \right],
\end{aligned}
\tag{3.58b}
$$

and the second order smearing function $f^{(2)}(X; t', t)$ is

$$
\begin{aligned}
f^{(2)}(X_1; t', t) &:= i^2 \operatorname{Re}\left[\langle \xi | \mu_D(t') [\mu_i(t), F(X_1)] | \xi \rangle \right] \\
&= -\eta_D(t')\eta_D(t) \operatorname{Re}\left[e^{-i\Omega\tau(t')} \langle 1 | [\mu_D(t), P_1] | 0 \rangle \right] \\
&= \eta_D(t')\eta_D(t) \operatorname{Re}\left[e^{-i\Omega[\tau(t') - \tau(t)]} \left(|\langle a | 1 \rangle|^2 \right) - |\langle a | 0 \rangle|^2 \right],
\end{aligned}
$$

$$\tag{3.59a}$$

$$
\begin{aligned}
f^{(2)}(X_2; t', t) &:= i^2 \operatorname{Re}\left[\langle \xi | \mu_D(t') [\mu_i(t), F(X_2)] | \xi \rangle \right] \\
&= -\eta_D(t')\eta_D(t) \operatorname{Re}\left[e^{-i\Omega[\tau(t') - \tau(t)]} \left(|\langle a | 1 \rangle|^2 \right) - |\langle a | 0 \rangle|^2 \right].
\end{aligned}
$$

$$\tag{3.59b}$$

Having computed these smearing functions, the POVM elements defining A are

$$A(X_1) = |\langle a|0\rangle|^2 I + 2\lambda \int dt\, \eta_D(t) \operatorname{Im}\left[e^{i\Omega\tau(t)} \langle 0|a\rangle \langle a|1\rangle\right]\phi[x_D(t)]$$

$$+ \lambda^2 \int dt'dt\, \eta_D(t')\eta_D(t) \operatorname{Re}\left[e^{-i\Omega[\tau(t')-\tau(t)]}\left(|\langle a|1\rangle|^2 - |\langle a|0\rangle|^2\right)\right]$$

$$\times \mathcal{T}\phi[x_D(t')]\phi[x_D(t)] + \mathcal{O}\left(\lambda^3\right),\tag{3.60a}$$

$$A(X_2) = I - A(X_1).\tag{3.60b}$$

Suppose we are interested in the probability that the detector has transitioned from its ground state $|0\rangle_D$ to its excited state $|1\rangle_D$ given the field is initially in the state $\rho \in \mathcal{S}\left(\mathcal{H}_\phi\right)$. This transition probability is given by $P_D = \operatorname{tr}(\rho A(X_1))$, with $|a\rangle_D = |1\rangle_D$, which defines the observable $F(X)$ in Eq. (3.56). Explicitly, this transition probability is given by

$$P_D = \lambda^2 \int dt'dt\, \eta_D(t')\eta_D(t) \operatorname{Re}\left[e^{-i\Omega[\tau(t')-\tau(t)]}\right]\langle\mathcal{T}\phi[x_D(t')]\phi[x_D(t)]\rangle_\rho$$

$$+ \mathcal{O}\left(\lambda^3\right)$$

$$= \lambda^2 \int dt'dt\, \eta_D(t')\eta_D(t)e^{-i\Omega[\tau(t')-\tau(t)]}\langle\phi[x_D(t')]\phi[x_D(t)]\rangle_\rho + \mathcal{O}\left(\lambda^3\right),$$

$$\tag{3.61}$$

where $\langle\,\cdot\,\rangle_\rho := \operatorname{tr}[\,\cdot\,\rho]$. When $\rho = |0\rangle_D\langle 0|_D \otimes |0\rangle\langle 0|$, with $|0\rangle \in \mathcal{H}_\phi$ being an appropriately defined vacuum state of the field, the transition probability P_D given in Eq. (3.61) reduces to the transition probability computed in Sect. 3.1 in Eq. (3.9).

3.3.2   Remarks and Future Applications

We close this section by making a few remarks on the measurement model presented above and comment on possible future applications.

1. Although effectively the same, an advantage of considering the field observables $A(X)$ rather than explicitly evolving a collection of detectors (as was done in Sect. 3.2) is that one clearly sees that the effect of changing the initial state of the detectors or the measured observable F is to change the smearing functions given in Eq. (3.53). In fact, all information about the detector model (size of Hilbert space, the operator $\mu(t)$, and the initial state of the detector) is encoded in the smearing function. This should allow for easy exploration of the effect of initially entangled detectors or detectors initially prepared in superpositions on various measurement tasks in quantum field theory because the only modification to the field observable $A(X)$ will be to the smearing functions.

Furthermore, constructing the observables a given detector model measures provides an algebraic picture in terms of the POVM elements $A(X)$ of possible physical measurements of a field. This may prove useful in making connections between the many studies of Unruh-DeWitt detectors and results in algebraic quantum field theory.

2. One may be able to construct an array of detectors prepared in such a way that the observables the detector model defines serve as an entanglement witness[7] for entanglement between different spacetime regions. This could lead to an operational formulation of the area law for entanglement in terms of a physical measurement model.

3. When studying the Unruh effect using uniformly accelerating Unruh-DeWitt detectors, one usually computes the transition probability P_D, which is second order in the interaction strength, and demonstrates that the final state of the detector is thermal

$$\rho_D = (1 - P_D)\,|0\rangle_D\langle 0|_D + P_D\,|1\rangle_D\langle 1|_D \in \mathcal{S}(\mathcal{K}),\qquad (3.62)$$

with a temperature $k_B T = \Omega/\ln[(1 - P_D)/P_D]$ proportional to the proper acceleration of the detector [9].

Instead, one may imagine preparing a collection of uniformly accelerating detectors in an appropriate state $|\xi\rangle \in \mathcal{K} = \bigotimes_i \mathcal{H}_i$ and performing a collective measurement (not a series of local measurements) of an observable F on the final state of the detectors, such that a signature of the Unruh effect appears at first order, rather than second order as in the approach described above. In essence, what one would be doing is exploring whether quantum phenomena, such as entanglement between detectors, can be used to create a better thermometer with which to measure the Unruh temperature, i.e. an observable F that is more sensitive to acceleration than a measurement of the transition probability.

4. Sorkin [42] has demonstrated that projective measurements of quantum fields can lead to superluminal signalling if consecutive projective measurements on the field have support on partial causally connected local regions of spacetime. This leads to the question: Which observables can be measured by a projective measurement in relativistic quantum field theory?

In general, the answer to this question is unknown. However, it is known that some observables, such as a Wilson loop in a non-abelian gauge theory [5] or a projector onto a one particle state [16, 42], cannot be measured.

To address this issue, Sorkin [42] called for a von Neumann-like analysis of the measurement process within quantum field theory. The above construction of detector observables, culminating in the POVM elements in Eq. (3.54), is such an analysis. It will be fruitful to examine these observables in relation to the issues raised by Sorkin.

---

[7]From [20]: A self-adjoint operator $W \in \mathcal{L}_s\,(\mathcal{H}_A \otimes \mathcal{H}_B)$ is an entanglement witness if W is not a positive operator but $\langle\psi|\,\langle\phi|\,W\,|\psi\rangle\,|\phi\rangle \geq 0$ for all factorized vectors $|\psi\rangle\,|\phi\rangle \in \mathcal{H}_A \otimes \mathcal{H}_B$. We say that an entangled state ρ is detected by W if $\mathrm{tr}[\rho W] < 0$.

3.4   Summary

We began this chapter in Sect. 3.1 by giving a physical motivation for the Unruh-DeWitt detector. We derived the probability that after the detector's interaction with the field it has transitioned from its ground state to its excited state, and the associated transition rate. The main results of this chapter were presented in Sects. 3.2 and 3.3.

In Sect. 3.2 we emphasized that the question of whether a quantum field is entangled is ultimately an operational one that depends on the measuring process and the motion of the observer making the measurement. We considered two initially unexcited Unruh-DeWitt detectors moving along arbitrary timelike trajectories in any curved spacetime admitting a Wightman function. We derived the final state ρ_{AB} of the detectors after their interaction with the field has ceased to all orders in the interaction strength λ, stating explicitly the leading order contribution to ρ_{AB} in terms of the Wightman function. We then computed the negativity, concurrence, entanglement of formation, and the correlation function between local measurements of each detector, which quantify the entanglement and correlations present in the state ρ_{AB}. We gave an interpretation of this entanglement as entanglement that has been transferred from the initial state of the field to these detectors, and interpreted these measures of entanglement as quantifying the entanglement between the regions in which the detectors were operating.

In Sect. 3.3 we showed that a collection of Unruh-DeWitt detectors constitutes a measurement model of an observable defined on the field Hilbert space $\mathcal{H}_{\phi}$. Through the probability reproducibility condition we explicitly derived the POVM elements associated with this observable to leading order and next to leading order in the interaction strength λ. Using these POVM elements we rederived the transition probability of a single detector. We also commented on possible applications of these observables.

References

1. J. Åke Larsson, Loopholes in Bell inequality tests of local realism. J. Phys. A **47**, 424003 (2014)
2. A.M. Alhambra, A. Kempf, E. Martín-Martínez, Casimir forces on atoms in optical cavities. Phys. Rev. A **89**, 033835 (2014)
3. M. Ali, A.R.P. Rau, G. Alber, Quantum discord for two-qubit X states. Phys. Rev. A **81**, 042105 (2010)
4. L. Amico, R. Fazio, A. Osterloh, V. Vedral, Entanglement in many-body systems. Rev. Mod. Phys. **80**, 517 (2008)
5. D. Beckman, D. Gottesman, A. Kitaev, J. Preskill, Measurability of Wilson loop operators. Phys. Rev. D **65**, 965022 (2002)
6. J.S. Bell, On the Einstein Podolsky and Rosen paradox. Physics **1**, 195 (1964)
7. J.S. Bell, On the problem of hidden variables in quantum mechanics. Rev. Mod. Phys. **38**, 447 (1966)

8. C.H. Bennett, D.P. DiVincenzo, J.A. Smolin, W.K. Wootters, Mixed-state entanglement and quantum error correction. Phys. Rev. A **54**, 3824 (1996)
9. N.D. Birrell, P.C.W. Davies, *Quantum Fields in Curved Space* (Cambridge University Press, Cambridge, 1982)
10. L. Bombelli, R.K. Koul, J. Lee, R.D. Sorkin, Quantum source of entropy for black holes. Phys. Rev. D **80**, 373 (1986)
11. D. Bruß, G. Leuchs, *Lectures on Quantum Information* (Wiley-VCH, Weinheim, 2007)
12. C. Callan, F. Wilczek, On geometric entropy. Phys. Lett. B **33**, 55 (1994)
13. H. Casini, M. Huerta, A finite entanglement entropy and the c-theorem. Phys. Lett. B **600**, 142 (2006)
14. H. Casini, M. Huerta, Entanglement entropy in free quantum field theory. J. Phys. A **41**, 504007 (2009)
15. B.S. DeWitt, Quantum gravity: the new synthesis, in *General Relativity: An Einstein Centenary Survey* (Cambridge University Press, Cambridge, 1979), pp. 680–745
16. F. Dowker, Useless qubits in "relativistic quantum information" (2011). arXiv:quant-ph/1111.2308
17. A. Einstein, B. Podolsky, N. Rosen, Can quantum mechanical description of physical reality be considered complete. Phys. Rev. Lett. **47**, 777 (1935)
18. C.J. Fewster, R. Verch, Algebraic quantum field theory in curved spacetimes, in *Advances in Algebraic Quantum Field Theory* (Springer, Cham, 2015), pp. 125–189
19. S.J. Freedman, J.F. Clauser, Experimental test of local hidden variable theories. Phys. Rev. Lett. **28**, 938 (1972)
20. T. Heinosaari, M. Ziman, *The Mathematical Language of Quantum Theory: From Uncertainty to Entanglement* (Cambridge University Press, Cambridge, 2012)
21. S. Hollands, R.M. Wald, Axiomatic quantum field theory in curved spacetime. Commun. Math. Phys. **293**, 85 (2010)
22. M. Horodecki, P. Horodecki, R. Horodecki, Separability of mixed states: necessary and sufficient conditions. Phys. Lett. A **223**, 1 (1996)
23. R. Horodecki, P. Horodecki, M. Horodecki, K. Horodecki, Quantum entanglement. Rev. Mod. Phys. **81**, 865 (2009)
24. J. Louko, A. Satz, Transition rate of the Unruh-DeWitt detector in curved spacetime. Classical Quantum Gravity **25**, 055012 (2008)
25. L. Mandel, E. Wolf, *Optical Coherence and Quantum Optics* (Cambridge University Press, Cambridge, 1995)
26. E. Martín-Martínez, E.G. Brown, W. Donnelly, A. Kempf, Sustainable entanglement production from a quantum field. Phys. Rev. A **88**, 052310 (2013)
27. E. Martín-Martínez, M. Montero, M. del Rey, Wavepacket detection with the Unruh-DeWitt model. Phys. Rev. D **87**, 064038 (2013)
28. E. Martín-Martínez, A.R.H. Smith, D.R. Terno, Spacetime structure and vacuum entanglement. Phys. Rev. D **93**, 044001 (2016)
29. M.A. Nielsen, I.L. Chuang, *Quantum Computation and Quantum Information* (Cambridge University Press, Cambridge, 2010)
30. A. Osterloh, L. Amico, G. Falci, R. Fazio, Scaling of entanglement close to a quantum phase transitions. Nature **416**, 608 (2002)
31. A. Peres, Separability criterion for density matrices. Phys. Rev. Lett. **77**, 1413 (1996)
32. A. Pozas-Kerstjens, E. Martín-Martínez, Harvesting correlations from the quantum vacuum. Phys. Rev. D **92**, 064042 (2015)
33. A. Pozas-Kerstjens, E. Martín-Martínez, Entanglement harvesting from the electromagnetic vacuum with hydrogenlike atoms. Phys. Rev. D **94**, 064074 (2016)
34. B. Reznik, Entanglement from the vacuum. Found. Phys. **33**, 167 (2003)
35. B. Reznik, A. Retzker, J. Silman, Violating Bell's inequalities in vacuum. Phys. Rev. A **71**, 042104 (2005)
36. S. Ryu, T. Takayanagi, Holographic derivation of entanglement entropy from AdS/CFT. Phys. Rev. Lett. **96**, 181602 (2006)

37. G. Salton, R.B. Mann, N.C. Menicucci, Acceleration-assisted entanglement harvesting and rangefinding. New J. Phys. **17**, 035001 (2015)
38. S. Schlicht, Considerations on the Unruh effect: causality and regularization. Class. Quant. Grav. **21**, 4647 (2004)
39. E. Schroödinger, Discussion of probability relations between separated systems. Math. Proc. Camb. Philos. Soc. **31**, 555–563 (1935)
40. M.O. Scully, M.S. Zubairy, *Quantum Optics* (Cambridge University Press, Cambridge, 1997)
41. J. Silman, B. Reznik, Three-region vacuum nonlocality (2005). arXiv:quant-ph/0501028
42. R.D. Sorkin, Impossible measurements on quantum fields, in *Directions in General Relativity*, ed. by B.L. Hu, T.A. Jacobson (Cambridge University Press, Cambridge, 1993), pp. 293–305
43. M. Srednicki, Entropy and area. Phys. Rev. Lett. **71**, 666 (1993)
44. S.J. Summers, R. Werner, The vacuum violates Bell's inequalities. Phys. Lett. A **110**, 257 (1985)
45. S.J. Summers, R. Werner, Bell's inequalities and quantum field theory. I. General setting. J. Math. Phys. **28**, 2440 (1987)
46. S.J. Summers, R. Werner, Bell's inequalities and quantum field theory. II. Bell's inequalities are maximally violated in the vacuum. J. Math. Phys. **28**, 2448 (1987)
47. G. Svetlichny, Distinguishing three-body from two-body nonseparability by a Bell-type inequality. Phys. Rev. D **35**, 3066 (1987)
48. W.G. Unruh, Notes on blackhole evaporation. Phys. Rev. D **14**, 870 (1976)
49. A. Valentini, Non-local correlations in quantum electrodynamics. Phys. Lett. A **153**, 321 (1991)
50. G. Ver Steeg, N.C. Menicucci, Entangling power of an expanding universe. Phys. Rev. D **79**, 044027 (2009)
51. G. Vidal, R.F. Werner, A computable measure of entanglement. Phys. Rev. A **65**, 032314 (2002)
52. G. Vidal, J.I. Latorre, E. Rico, A. Kitaev, Entanglement in quantum critical phenomena. Phys. Rev. Lett. **90**, 227902 (2003)
53. C. Weedbrook, S. Pirandola, R. García-Patrón, N.J. Cerf, T.C. Ralph, J.H. Shapiro, S. Lloyd, Gaussian quantum information. Rev. Mod. Phys. **84**, 621 (2012)
54. W.K. Wootters, Entanglement of formation of an arbitrary state of two qubits. Phys. Rev. Lett. **80**, 2245 (1998)

Chapter 4
Unruh-DeWitt Detectors in Quotients of Minkowski Space

We now apply the formalism developed in Chap. 3 to study the behaviour of Unruh-DeWitt detectors in Minkowski space $\mathcal{M}$ as compared to two distinct cylindrical spacetimes, which we will refer to as $\mathcal{M}_0$ and $\mathcal{M}_-$, constructed by topological identifications of Minkowski space. These identifications are implemented by quotienting Minkowski space with an appropriate group; spacetimes constructed this way are referred to as quotient spacetimes, a general discussion of which is given in Appendix B.1.

The purpose of this chapter is to investigate how the transition probability of a single Unruh-DeWitt detector and the entanglement and correlations harvested by two such detectors are affected by topological identifications. We begin in Sect. 4.1 by constructing the quotient spacetimes $\mathcal{M}_0$ and $\mathcal{M}_-$ and derive the Wightman function associated with the vacuum state of twisted and untwisted real scalar fields in each via the method of images. In Sect. 4.2 we compute the transition probability of a single inertial detector in all three spacetimes $\mathcal{M}$, $\mathcal{M}_0$, and $\mathcal{M}_-$. In Sect. 4.3 we compare the amount of entanglement harvested from the vacuum state by two detectors in all three spacetimes, and demonstrate that the orientation of detectors with respect to a topological identification affects how much entanglement the detectors can harvest. The conclusion of this investigation is that, as seen by detectors, the entanglement structure of the vacuum state is affected by the topology of spacetime.

4.1 Quotients of Minkowski Space and Their Wightman Functions

The quotient spacetimes $\mathcal{M}_0$ and $\mathcal{M}_-$ are constructed by making topological identifications of Minkowski space. These identifications will be expressed in Minkowski coordinates t, x, y, and z, in which the Minkowski line element takes

© Springer Nature Switzerland AG 2019
A. R. H. Smith, *Detectors, Reference Frames, and Time*, Springer Theses,
https://doi.org/10.1007/978-3-030-11000-0_4

the familiar form

$$ds^2 = dt^2 - dx^2 - dy^2 - dz^2. \tag{4.1}$$

The first quotient spacetime we consider is the cylindrical spacetime

$$\mathcal{M}_0 := \mathcal{M}/Z, \tag{4.2}$$

which is constructed by quotienting Minkowski space $\mathcal{M}$ with the group $Z \simeq \{J_0^n\}$, where the generator of the group J_0 acts on $\mathcal{M}$ by identifying points

$$J_0 : (t, x, y, z) \sim (t, x, y, z + a), \tag{4.3}$$

where a is the circumference of spacetime.

The second quotient spacetime we consider is

$$\mathcal{M}_- := \mathcal{M}/\Gamma_-, \tag{4.4}$$

where the group $\Gamma_- \simeq \{J_-^n\}$ is generated by the discrete isometry

$$J_- : (t, x, y, z) \sim (t, -x, -y, z + a), \tag{4.5}$$

where again a is the circumference of the universe. $\mathcal{M}_-$ is a cylindrical spacetime in which rotations by π in the xy-plane have been identified.

Both identifications, J_0 and J_-, preserve space and time orientation and act freely and properly ensuring both $\mathcal{M}_0$ and $\mathcal{M}_-$ are space and time orientable Lorentzian manifolds [4]. As neither J_0 nor J_- affect the Minkowski line element, both $\mathcal{M}_0$ and $\mathcal{M}_-$ are locally flat spacetimes.

To study the behaviour of Unruh-DeWitt detectors in $\mathcal{M}_0$ and $\mathcal{M}_-$ we need to compute the final state of the detectors after interacting with a quantum field. For a single detector initially unexcited, this amounts to computing the transition probability P_D given in Eq. (3.9). For two detectors, both initially unexcited, in addition to the transition probability of each detector, one must compute the matrix elements X and C appearing in the final state of the two detectors given in Eqs. (3.24). Assuming the state of the field is prepared in its vacuum state,[1] all of these quantities are given by integrations over the Wightman function associated with the vacuum state of the field.

---

[1]In this chapter, we will simply say 'vacuum state' without explicitly stating the vacuum state agreed on by all inertial observers.

Constructing Wightman functions associated with quantum fields on quotient spacetimes is done using the method of images. Given that one knows the Wightman function in the spacetime being identified (in the present case this is Minkowski space $\mathcal{M}$), the Wightman function in the quotient spacetime $\mathcal{M}/G$ is given by the image sum

$$W_{\mathcal{M}/G}(x, x') = \sum_n \eta^n \, W_{\mathcal{M}}(x, g^n x'), \tag{4.6}$$

where G is the group of identifications acting on $\mathcal{M}$, $g \in G$ is the generator of the group, $g^n x'$ denotes the action of the group element g^n on the spacetime point x'—the group action being realized by an identification associated with g^n, and $W_{\mathcal{M}}(x, x')$ and $W_{\mathcal{M}/G}(x, x')$ are the Wightman functions in $\mathcal{M}$ and the quotient spacetime $\mathcal{M}/G$; the summation is carried out over all elements of G. The parameter η is equal to -1 for twisted fields and 1 for untwisted fields. The method of images applied to the study of Wightman functions (and other Green's functions) was first investigated by Banach and Dowker [1, 2]. An expanded discussion of quotient spacetimes and the method of images is given in Appendix B.1.

4.1.1   The Wightman Function in Minkowski Space

To compute the Wightman function in $\mathcal{M}_0$ and $\mathcal{M}_-$ via the method of images we require the Wightman function in Minkowski space, which we will now derive. We restrict our attention to massless fields in (3+1)-dimensional Minkowski space. The equation of motion satisfied by a free scalar field, Eq. (2.4), simplifies in this case to

$$\partial^\mu \partial_\mu \, \phi(x) = 0. \tag{4.7}$$

A complete orthonormal set of mode functions with respect to the inner product in Eq. (2.5) is given by the plane wave solutions

$$u_{\mathbf{k}}(x) = \frac{1}{\sqrt{2\,|\mathbf{k}|\,(2\pi)^3}} e^{-i|\mathbf{k}|t + i\mathbf{k}\cdot\mathbf{x}}, \tag{4.8}$$

where $x = (t, \mathbf{x})$. The field may be expanded in terms of these mode functions as

$$\phi(x) = \int \frac{dk^3}{(2\pi)^{3/2}} \frac{1}{\sqrt{2\,|\mathbf{k}|}} \left(e^{-i|\mathbf{k}|t + i\mathbf{k}\cdot\mathbf{x}} a_{\mathbf{k}} + e^{i|\mathbf{k}|t - i\mathbf{k}\cdot\mathbf{x}} a_{\mathbf{k}}^\dagger \right). \tag{4.9}$$

Once the theory is quantized, the operators $a_{\mathbf{k}}$ and $a_{\mathbf{k}}^\dagger$ satisfy the commutation relations in Eq. (2.11).

Letting $|0\rangle$ denote the Minkowski vacuum state and using the expansion of the field in Eq. (4.9), the Wightman function may be computed

$$W_{\mathcal{M}}(x, x') := \langle 0|\phi(x)\phi(x')|0 \rangle$$

$$= \frac{1}{(2\pi)^3} \int \frac{dk^3}{2\,|\mathbf{k}|}\, e^{i\mathbf{k}\cdot(\mathbf{x}-\mathbf{x}')-i|\mathbf{k}|(t-t')}$$

$$= \frac{1}{4\pi^2\,|\mathbf{x}-\mathbf{x}'|} \int_0^\infty d|\mathbf{k}|\, e^{-i|\mathbf{k}|(t-t')} \sin\big[|\mathbf{k}|\,|\mathbf{x}-\mathbf{x}'|\big]$$

$$= \frac{1}{4\pi^2\,|\mathbf{x}-\mathbf{x}'|} \lim_{\epsilon \to 0^+} \int_0^\infty d|\mathbf{k}|\, e^{-i|\mathbf{k}|[(t-t')-i\epsilon\,\mathrm{sgn}(t-t')]} \sin\big[|\mathbf{k}|\,|\mathbf{x}-\mathbf{x}'|\big)\big]$$

$$= -\frac{1}{4\pi^2} \lim_{\epsilon \to 0^+} \frac{1}{\big[t-t'-i\epsilon\,\mathrm{sgn}(t-t')\big]^2 - \mathbf{x}^2}$$

$$= \frac{1}{4\pi i}\,\mathrm{sgn}(t-t')\,\delta\big[\sigma(x,x')\big] - \frac{1}{4\pi^2\sigma(x,x')}, \tag{4.10}$$

where

$$\sigma(x, x') := (t-t')^2 - (x-x')^2 - (y-y')^2 - (z-z')^2, \tag{4.11}$$

and in arriving at the last equality we made use of Sokhotsky's formula.[2]

4.1.2   The Wightman Function in $\mathcal{M}_0$ and $\mathcal{M}_-$

Having derived the Wightman function in Minkowski space, we can compute the Wightman function in both $\mathcal{M}_0$ and $\mathcal{M}_-$ using the method of images.

In $\mathcal{M}_0$, the Wightman function is

$$W_{\mathcal{M}_0}(x, x') = \sum_{n=-\infty}^{\infty} \eta^n\, W_M(x, J_0^n x')$$

$$= \sum_{n=-\infty}^{\infty} \eta^n \left[\frac{1}{4\pi i}\,\mathrm{sgn}(t-t')\,\delta\big[\sigma(x, J_0^n x')\big] - \frac{1}{4\pi^2\sigma(x, J_0^n x')} \right], \tag{4.12}$$

where

$$\sigma(x, J_0^n x') = (t-t')^2 - (x-x')^2 - (y-y')^2 - (z-z'-an)^2. \tag{4.13}$$

---

[2] $\lim_{\epsilon \to 0} \frac{1}{x \pm i\epsilon} = \mp i\pi\delta(x) + \mathrm{PV}\,\frac{1}{x}$.

Similarly, the Wightman function in $\mathcal{M}_-$ is

$$W_{\mathcal{M}_-}(x, x') = \sum_{n=-\infty}^{\infty} \eta^n \, W_M(x, J_-^n x')$$

$$= \sum_{n=-\infty}^{\infty} \eta^n \left[\frac{1}{4\pi i} \, \mathrm{sgn}(t - t') \, \delta[\sigma(x, J_-^n x')] - \frac{1}{4\pi^2 \sigma(x, J_-^n x')} \right],$$

(4.14)

where

$$\sigma(x, J_-^n x') = (t - t')^2 - (x - (-1)^n x')^2 - (y - (-1)^n y')^2 - (z - z' - an)^2.$$

(4.15)

4.2   The Transition Probability in $\mathcal{M}$, $\mathcal{M}_0$, and $\mathcal{M}_-$

Having constructed the Wightman function in $\mathcal{M}$, $\mathcal{M}_0$, and $\mathcal{M}_-$, we now turn our attention to the behaviour of Unruh-DeWitt detectors in these spacetimes. Unruh-DeWitt detectors in $\mathcal{M}_0$ and $\mathcal{M}_-$ were first investigated by Langlois [4], who computed the transition rate of inertial and uniformly accelerating detectors. In this section we will compare the transition probability of a single detector in all three spacetimes coupled to both twisted and untwisted fields.

Suppose a detector is initially prepared in its ground state and remains at rest with respect to the coordinate system (t, x, y, z) throughout the interaction with the scalar field with its trajectory given by

$$x_D(\tau) := \{t = \tau, \; \mathbf{x} = (d_x, d_y, d_z)\}.$$

(4.16)

We choose the switching function $\chi_D(\tau)$ controlling the duration of the interaction with the scalar field to be a Gaussian

$$\chi_D(\tau) = e^{-\tau^2/2\sigma^2},$$

(4.17)

with the interpretation that the detector is interacting with the field for an approximate amount of proper time $k\sigma$, where $k \in \mathbb{R}$ is chosen so that the interaction between the detector and field is negligible at the proper time $\tau = k\sigma$.

With these choices, the transition probability as given in Eq. (3.9) in either $\mathcal{M}$, $\mathcal{M}_0$, or $\mathcal{M}_-$ simplifies to

$$P_D = \lambda^2 \int d\tau d\tau' \, e^{-\tau^2/2\sigma^2} e^{-\tau'^2/2\sigma^2} e^{-i\Omega(\tau-\tau')} W\big(x_D(\tau), x_D(\tau')\big) + \mathcal{O}\big(\lambda^4\big),$$

(4.18)

where $W\big(x_D(\tau), x_D(\tau')\big)$ is the Wightman function in either $\mathcal{M}$, $\mathcal{M}_0$, or $\mathcal{M}_-$ evaluated along the detector's trajectory $x_D(\tau)$.

We will now evaluate Eq. (4.18) for detectors in $\mathcal{M}$, $\mathcal{M}_0$, or $\mathcal{M}_-$. The reader who is uninterested in an explicit evaluation of Eq. (4.18) for the Wightman functions given in Eqs. (4.10), (4.12), and (4.14) may skip to Table 4.1, where the transition probabilities in all three spacetimes are summarized.

4.2.1 The Transition Probability in Minkowski Space

We first calculate the transition probability in Minkowski space, which we will denote as $P_{\mathcal{M}}$. Substituting the Minkowski space Wightman function, Eq. (4.10), evaluated along the detector's trajectory given in Eq. (4.16) into Eq. (4.18) results in

$$P_{\mathcal{M}} = \lambda^2 \int_{-\infty}^{\infty} d\tau \int_{-\infty}^{\infty} d\tau' \, e^{-(\tau+\tau')^2/4\sigma^2} e^{-(\tau-\tau')^2/4\sigma^2} e^{-i\Omega(\tau-\tau')}$$
$$\times \left[\frac{1}{4\pi i} \operatorname{sgn}(\tau - \tau') \delta\big((\tau - \tau')^2\big) - \frac{1}{4\pi^2(\tau-\tau')^2} \right] + \mathcal{O}\big(\lambda^4\big). \quad (4.19)$$

Let us change the integration variables to

$$y := \tau - \tau' \quad \text{and} \quad y' := \tau + \tau', \quad (4.20)$$

noting that the volume element transforms as $d\tau d\tau' = \frac{1}{2} dy dy'$. With these variables the transition probability becomes

$$P_{\mathcal{M}} = \frac{\lambda^2}{2} \int_{-\infty}^{\infty} dy \int_{-\infty}^{\infty} dy' \, e^{-y'^2/4\sigma^2} e^{-y^2/4\sigma^2} e^{-i\Omega y}$$
$$\times \left[\frac{1}{4\pi i} \operatorname{sgn}(y) \delta\big(y^2\big) - \frac{1}{4\pi^2 y^2} \right] + \mathcal{O}\big(\lambda^4\big)$$
$$= \lambda^2 \sqrt{\pi} \sigma \int_{-\infty}^{\infty} dy \, e^{-y^2/4\sigma^2} e^{-i\Omega y} \left[\frac{1}{4\pi i} \operatorname{sgn}(y) \delta\big(y^2\big) - \frac{1}{4\pi^2 y^2} \right] + \mathcal{O}\big(\lambda^4\big)$$
$$= \frac{\lambda^2}{4\pi} \sqrt{\pi} \sigma \, [T_1 + T_2] + \mathcal{O}\big(\lambda^4\big), \quad (4.21)$$

where in the last equality

$$T_1 := -i \int_{-\infty}^{\infty} dy \, e^{-y^2/4\sigma^2} e^{-i\Omega y} \operatorname{sgn}(y) \delta\big(y^2\big), \quad (4.22a)$$

$$T_2 := -\frac{1}{\pi} \int_{-\infty}^{\infty} dy \, e^{-y^2/4\sigma^2} e^{-i\Omega y} \frac{1}{y^2}. \quad (4.22b)$$

Let us begin with the evaluation of T_1. Observe that for a well-behaved[3] test function $f(y)$, we have

$$\text{PV} \int_{-\infty}^{\infty} dy \, f(y) \, \text{sgn}(y) \, \delta\left(y^2\right)$$

$$= \lim_{r\to 0} \text{PV} \int_{-\infty}^{\infty} dy \, f(y) \, \text{sgn}(y) \, \delta\left(y^2 - r^2\right)$$

$$= \lim_{r\to 0} \text{PV} \int_{-\infty}^{\infty} dy \, f(y) \, \text{sgn}(y) \, \frac{1}{2\,|r|} \left[\delta(y - r) + \delta(y + r)\right]$$

$$= \lim_{r\to 0} \frac{f(r) - f(-r)}{2\,|r|}$$

$$=: f'(0). \tag{4.23}$$

Using this result, T_1 simplifies to

$$T_1 = -i \, \frac{d}{dy} e^{-y^2/4\sigma^2} e^{-i\Omega y} \Big|_{y=0} = -\Omega. \tag{4.24}$$

To evaluate T_2, we make use of the identity

$$\text{PV} \int_{-\infty}^{\infty} dy \, \frac{f(y)}{y^2} = \int_{0}^{\infty} dy \, \frac{f(y) + f(-y) - 2f(0)}{y^2}, \tag{4.25}$$

which is derived in Appendix B.2. Using Eq. (4.25), T_2 may be evaluated

$$T_2 = -\frac{1}{\pi} \int_{0}^{\infty} dy \, \frac{e^{-y^2/4\sigma^2} e^{-i\Omega y} + e^{-y^2/4\sigma^2} e^{i\Omega y} - 2}{y^2}$$

$$= -\frac{2}{\pi} \int_{0}^{\infty} dy \, \frac{e^{-y^2/4\sigma^2} \cos(\Omega y) - 1}{y^2}$$

$$= -\frac{2}{\pi} \left[-\frac{\sqrt{\pi}}{2\sigma} \left(e^{-\sigma^2 \Omega^2} + \sqrt{\pi} \sigma \Omega \, \text{erf}(\sigma \Omega) \right) \right]$$

$$= \frac{1}{\sqrt{\pi}\sigma} e^{-\sigma^2 \Omega^2} + \Omega \, \text{erf}(\sigma \Omega). \tag{4.26}$$

Having calculated both T_1 and T_2, using Eq. (4.21) the transition probability is given by

$$P_{\mathcal{M}} = \frac{\lambda^2}{4\pi} \left[e^{-\sigma^2 \Omega^2} - \sqrt{\pi}\sigma \Omega \, \text{erfc}(\sigma \Omega) \right] + \mathcal{O}\left(\lambda^4\right). \tag{4.27}$$

---

[3] A smooth function that tends to zero as $y \to \pm\infty$.

4.2.2   The Transition Probability in $\mathcal{M}_0$

We now compute the transition probability for the same detector considered above in the cylindrical spacetime $\mathcal{M}_0$, which we will label by $P_{\mathcal{M}_0}$. Again, beginning with the expression for the transition probability given in Eq. (4.18) and making use of the Wightman function in Eq. (4.12) evaluated along the trajectory in Eq. (4.16), the transition probability of a detector in $\mathcal{M}_0$ is given by

$$P_{\mathcal{M}_0} = P_{\mathcal{M}} + \lambda^2 \sum_{n \neq 0} \eta^n \int_{-\infty}^{\infty} d\tau \int_{-\infty}^{\infty} d\tau' \, e^{-(\tau+\tau')^2/4\sigma^2} e^{-(\tau-\tau')^2/4\sigma^2} e^{-i\Omega(\tau-\tau')}$$

$$\times \left[\frac{1}{4\pi i} \operatorname{sgn}(\tau - \tau') \, \delta\left((\tau - \tau')^2 - a^2 n^2\right) - \frac{1}{4\pi^2 \left[(\tau - \tau')^2 - a^2 n^2\right]} \right]$$

$$+ \mathcal{O}\left(\lambda^4\right)$$

$$= P_{\mathcal{M}} + \lambda^2 \sqrt{\pi} \sigma \sum_{n \neq 0} \eta^n \int_{-\infty}^{\infty} dy \, e^{-y^2/4\sigma^2} e^{-i\Omega y}$$

$$\times \left[\frac{1}{4\pi i} \operatorname{sgn}(y) \, \delta\left(y^2 - a^2 n^2\right) - \frac{1}{4\pi^2 \left[y^2 - a^2 n^2\right]} \right] + \mathcal{O}\left(\lambda^4\right),$$

$$\tag{4.28}$$

where the last equality is obtained by changing the integration variables to $y := \tau - \tau'$ and $y' := \tau + \tau'$ and carrying out the integration over y'. Defining

$$T_1(n) := \frac{1}{4\pi i} \int_{-\infty}^{\infty} dy \, e^{-y^2/4\sigma^2} e^{-i\Omega y} \operatorname{sgn}(y) \, \delta\left(y^2 - a^2 n^2\right), \tag{4.29a}$$

$$T_2(n) := -\frac{1}{4\pi^2} \int_{-\infty}^{\infty} dy \, e^{-y^2/4\sigma^2} e^{-i\Omega y} \frac{1}{y^2 - a^2 n^2}, \tag{4.29b}$$

allows the transition probability to be written as

$$A_{\mathcal{M}_0} = A_{\mathcal{M}} + \lambda^2 2\sqrt{\pi} \sigma \sum_{n=1}^{\infty} \eta^n \left[T_1(n) + T_2(n)\right] + \mathcal{O}\left(\lambda^4\right). \tag{4.30}$$

The integration in $T_1(n)$ may be carried out, yielding

$$T_1(n) = \frac{1}{4\pi i} \int_{-\infty}^{\infty} dy \, e^{-y^2/4\sigma^2} e^{-i\Omega y} \operatorname{sgn}(y) \, \delta\left(y^2 - a^2 n^2\right)$$

$$= \frac{1}{4\pi i} \int_{-\infty}^{\infty} dy \, e^{-y^2/4\sigma^2} e^{-i\Omega y} \operatorname{sgn}(y) \frac{1}{2|an|} \left[\delta(y + an) + \delta(y - an)\right]$$

$$= \frac{1}{4\pi} e^{-a^2n^2/4\sigma^2} \frac{1}{|an|} \frac{\text{sgn}(an)e^{-i\Omega an} + \text{sgn}(-an)e^{i\Omega an}}{2i}$$

$$= \frac{1}{4\pi} e^{-a^2n^2/4\sigma^2} \frac{\text{sgn}(an)}{|an|} \sin(\Omega an). \tag{4.31}$$

The integration in $T_2(n)$ may also be performed, yielding

$$T_2(n) = -\frac{1}{4\pi^2} \int_{-\infty}^{\infty} dy\, e^{-y^2/4\sigma^2} e^{-i\Omega y} \frac{1}{y^2 - a^2n^2}$$

$$= -\frac{1}{4\pi^2} \int_{-\infty}^{\infty} dy\, d\bar{y}\, \delta(\bar{y} - y)\, e^{-\bar{y}^2/4\sigma^2} e^{-i\Omega \bar{y}} \frac{1}{y^2 - a^2n^2}$$

$$= -\frac{1}{4\pi^2} \int_{-\infty}^{\infty} dy\, d\bar{y} \left(\frac{1}{2\pi} \int_{-\infty}^{\infty} dz\, e^{iz(\bar{y}-y)} \right) e^{-\bar{y}^2/4\sigma^2} e^{-i\Omega \bar{y}} \frac{1}{y^2 - a^2n^2}$$

$$= -\frac{1}{8\pi^3} \int_{-\infty}^{\infty} dz \left(\int_{-\infty}^{\infty} d\bar{y}\, e^{-i(\Omega-z)\bar{y}} e^{-\bar{y}^2/4\sigma^2} \right)$$

$$\times \left(\int_{-\infty}^{\infty} dy\, e^{-izy} \frac{1}{y^2 - a^2n^2} \right). \tag{4.32}$$

The integrals in the round brackets above are recognized as Fourier transforms of $e^{-\bar{y}^2/4\sigma^2}$ and $1/(t^2 - a^2n^2)$, which may be evaluated using [3] to give

$$T_2(n) = -\frac{1}{8\pi^3} \int_{-\infty}^{\infty} dz \left(2\sqrt{\pi}\sigma e^{-(\Omega-z)^2\sigma^2} \right) \left(-\pi \,\text{sgn}(z) \frac{\sin(anz)}{an} \right)$$

$$= \frac{\sigma}{4\pi^{3/2}an} \int_{-\infty}^{\infty} dz\, \text{sgn}(z) \sin(anz) e^{-(\Omega-z)^2\sigma^2}$$

$$= \frac{\sigma}{4\pi\,an} e^{-a^2n^2/4\sigma^2} \,\text{Im}\!\left[e^{ian\Omega} \,\text{erf}\!\left(i\frac{an}{2\sigma} + \sigma\Omega \right) \right], \tag{4.33}$$

where the integration over z was performed with Mathematica and $\text{erf}(x) := \frac{2}{\sqrt{\pi}} \int_0^x dx'\, e^{-x'^2}$ is the error function.

Having evaluated both $T_1(n)$ and $T_2(n)$, the transition probability in $\mathcal{M}_0$ simplifies to

$$P_{\mathcal{M}_0} = P_{\mathcal{M}}$$

$$+ \frac{\lambda^2\sigma}{4\sqrt{\pi}} \sum_{n=1}^{\infty} \eta^n \frac{e^{-a^2n^2/4\sigma^2}}{an} \left(\text{Im}\!\left[e^{ian\Omega} \,\text{erf}\!\left(i\frac{an}{2\sigma} + \sigma\Omega \right) \right] - \sin(\Omega an) \right)$$

$$+ \mathcal{O}\!\left(\lambda^4 \right). \tag{4.34}$$

4.2.3   The Transition Probability in $\mathcal{M}_-$

We now compute the transition probability of the same detector considered above in the cylindrical spacetime $\mathcal{M}_-$, which we will denote as $P_{\mathcal{M}_-}$. Again, we begin with the expression for the transition probability given in Eq. (4.18), and substitute the Wightman function given in Eq. (4.14) evaluated along the detector's trajectory specified in Eq. (4.16), which results in

$$
P_{\mathcal{M}_-} = A_{\mathcal{M}} + \lambda^2 \sum_{n \neq 0} \eta^n \int_{-\infty}^{\infty} d\tau \int_{-\infty}^{\infty} d\tau' \, e^{-(\tau+\tau')^2/4\sigma^2} e^{-(\tau-\tau')^2/4\sigma^2} e^{-i\Omega(\tau-\tau')}
$$

$$
\times \left[\frac{1}{4\pi i} \operatorname{sgn}(\tau - \tau') \, \delta\!\left((\tau - \tau')^2 - (d_x^2 + d_y^2)(1 - (-1)^n) - a^2 n^2 \right) \right.
$$

$$
\left. - \frac{1}{4\pi^2 \, [(\tau - \tau')^2 - (d_x^2 + d_y^2)(1 - (-1)^n) - a^2 n^2]} \right] + \mathcal{O}\!\left(\lambda^4 \right)
$$

$$
= A_{\mathcal{M}} + \lambda^2 2\sqrt{\pi}\sigma \sum_{n=1}^{\infty} \eta^n \int_{-\infty}^{\infty} dy \, e^{-y^2/4\sigma^2} e^{-i\Omega y}
$$

$$
\times \left[\frac{1}{4\pi i} \operatorname{sgn}(y) \, \delta\!\left(y^2 - D(n)^2 \right) - \frac{1}{4\pi^2 [y^2 - D(n)^2]} \right] + \mathcal{O}\!\left(\lambda^4 \right),
$$

$$
\tag{4.35}
$$

where

$$
D(n)^2 := \left[1 - (-1)^n \right] \left(d_x^2 + d_y^2 \right) + a^2 n^2,
\tag{4.36}
$$

and the last equality is obtained by noting the sum is invariant under $n \to -n$, and changing the integration variables to $y := \tau - \tau'$ and $y' := \tau + \tau'$, and carrying out the integration over y'.

Upon comparison of Eq. (4.35) to Eq. (4.28), we see that the transition probability in $\mathcal{M}_-$ is identical to $\mathcal{M}_0$ under the replacement of $an \to D(n)$. Thus, the transition probability in $\mathcal{M}_-$ is given by

$$
P_{\mathcal{M}_-} = P_{\mathcal{M}} + \frac{\lambda^2 \sigma}{2\sqrt{\pi}} \sum_{n=1}^{\infty} \eta^n \frac{e^{-D(n)^2/4\sigma^2}}{D(n)} \left(\operatorname{Im}\!\left[e^{i D(n)\Omega} \operatorname{erf}\!\left(\frac{i D(n)}{2\sigma} + \sigma\Omega \right) \right] \right.
$$

$$
\left. - \sin(D(n)\Omega) \right) + \mathcal{O}\!\left(\lambda^4 \right).
\tag{4.37}
$$

4.2.4 Comparison of the Transition Probability in $\mathcal{M}$, $\mathcal{M}_0$, and $\mathcal{M}_-$

The transition probabilities of an Unruh-DeWitt detector in $\mathcal{M}$, $\mathcal{M}_0$, and $\mathcal{M}_-$, as calculated above, are summarized in Table 4.1. We compare these probabilities by plotting them as a function of the energy gap of the detector $\sigma\Omega$ in Figs. 4.1, 4.2, and 4.3. Note that a negative energy gap $\sigma\Omega < 0$ corresponds to the de-excitation probability of a detector, i.e. the probability that if the detector began in the excited state $|1\rangle_D$ it has transitioned to its ground state $|0\rangle_D$ after the interaction with the field has ceased. For plotting purposes the image sums appearing in the transition probabilities for detectors in $\mathcal{M}_0$ and $\mathcal{M}_-$, Eqs. (4.34) and (4.37), have been truncated after 100 terms, which results in an error on the order of 10^{-5}.

From Figs. 4.1, 4.2, and 4.3 we make the following observations:

1. When the circumference a/σ of either cylindrical spacetime, $\mathcal{M}_0$ or $\mathcal{M}_-$, becomes large, the transition probability of a detector in both spacetimes approaches the transition probability of an equivalent detector in Minkowski space. For smaller circumferences, the transition probability for negative energy gaps oscillates around the transition probability of an equivalent detector in Minkowski space as a function of the detectors energy gap $\sigma\Omega$, with the frequency increasing as the circumference of either $\mathcal{M}_0$ or $\mathcal{M}_-$ increases.
2. From Fig. 4.1, we see that the difference between a detector coupled to (a) an untwisted field versus (b) a twisted field in $\mathcal{M}_0$ is that the oscillations of the transition probability as a function of the detector's energy gap for negative

Table 4.1 The transition probability in Minkowski space $\mathcal{M}$ and the two cylindrical spacetimes $\mathcal{M}_0$ and $\mathcal{M}_-$ constructed by the identifications $(t, x, y, z) \sim (t, x, y, z + a)$ and $(t, x, y, z) \sim (t, -x, -y, z + a)$ of Minkowski space, respectively

Spacetime	Transition probability to leading order in λ
$\mathcal{M}$	$P_{\mathcal{M}} = \dfrac{\lambda^2}{4\pi}\left[e^{-\sigma^2\Omega^2} - \sqrt{\pi}\,\sigma\Omega\,\mathrm{erfc}(\sigma\Omega)\right]$
$\mathcal{M}_0$	$P_{\mathcal{M}_0} = P_{\mathcal{M}} + \dfrac{\lambda^2\sigma}{2\sqrt{\pi}}\sum_{n=1}^{\infty}\eta^n\dfrac{e^{-a^2n^2/4\sigma^2}}{an}$ $\times\left(\mathrm{Im}\left[e^{ian\Omega}\,\mathrm{erf}\left(i\dfrac{an}{2\sigma} + \sigma\Omega\right)\right] - \sin(\Omega an)\right)$
$\mathcal{M}_-$	$P_{\mathcal{M}_-} = P_{\mathcal{M}} + \dfrac{\lambda^2\sigma}{2\sqrt{\pi}}\sum_{n=1}^{\infty}\eta^n\dfrac{e^{-D(n)^2/4\sigma^2}}{D(n)}$ $\times\left(\mathrm{Im}\left[e^{iD(n)\Omega}\,\mathrm{erf}\left(i\dfrac{D(n)}{2\sigma} + \sigma\Omega\right)\right] - \sin(\Omega D(n))\right)$

Note that $D(n)^2 := [1 - (-1)^n]\left(d_x^2 + d_y^2\right) + a^2 n^2$

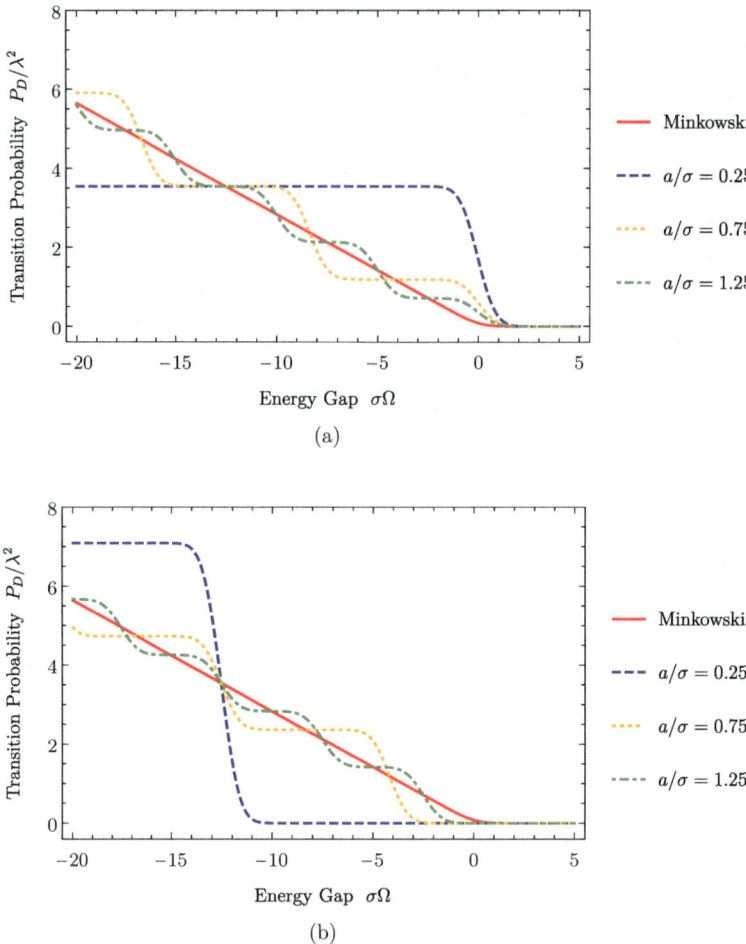

Fig. 4.1 The transition probability of a detector coupled to an (**a**) untwisted ($\eta = 1$) and (**b**) twisted ($\eta = -1$) massless scalar field in Minkowski space $\mathcal{M}$ (solid red) is compared to the transition probability of the same detector in the cylindrical spacetime $\mathcal{M}_0$ (broken lines) by plotting the transition probability as a function of the energy gap $\sigma\Omega$ of the detector. Different circumferences a/σ of the cylindrical spacetime are shown

energy gap are exactly out of phase. Further, the transition probability of a detector with a positive energy gap coupled to an untwisted field is significantly larger than the same detector coupled to a twisted field.

3. From Figs. 4.2 and 4.3, it is seen that the distance $\sqrt{d_x^2 + d_y^2}/\sigma$ the detector is away from the origin in the xy-plane in $\mathcal{M}_-$ affects the transition probability. For a detector located at the origin of the xy-plane, the transition probability of a detector in $\mathcal{M}_-$ is identical to a detector in $\mathcal{M}_0$ with the same circumference. As the detector moves away from the origin in the xy-plane, additional oscillations

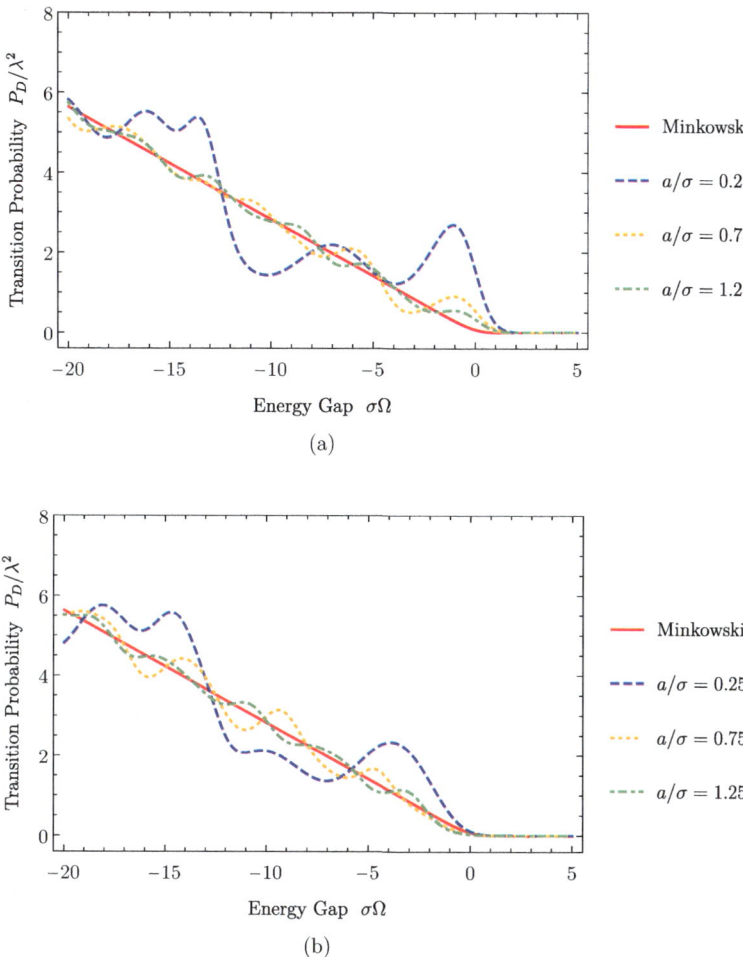

Fig. 4.2 The transition probability of a detector coupled to an (**a**) untwisted ($\eta = 1$) and (**b**) twisted ($\eta = -1$) massless scalar field in Minkowski space $\mathcal{M}$ (solid red) is compared to the transition probability of the same detector in the cylindrical spacetime $\mathcal{M}_-$ (broken lines) by plotting the transition probability as a function of the energy gap $\sigma\Omega$ of the detector. The detector is located a distance $\sqrt{d_x^2 + d_y^2}/\sigma = 0.5$ away from the origin in the xy-plane. Different circumferences a/σ of the cylindrical spacetime are shown

appear in the transition probability (Fig. 4.2) and then disappear when the detector is far from the origin in the xy-plane (Fig. 4.3).

4. From Figs. 4.2 and 4.3, the difference in the transition probability of a detector coupled to an untwisted versus twisted field in $\mathcal{M}_-$ only appears when the detector is close to the origin in the xy-plane (Fig. 4.2). As the detector moves away from the origin in the xy-plane (Fig. 4.3), the transition probability is insensitive to whether the detector is coupled to a twisted or untwisted field.

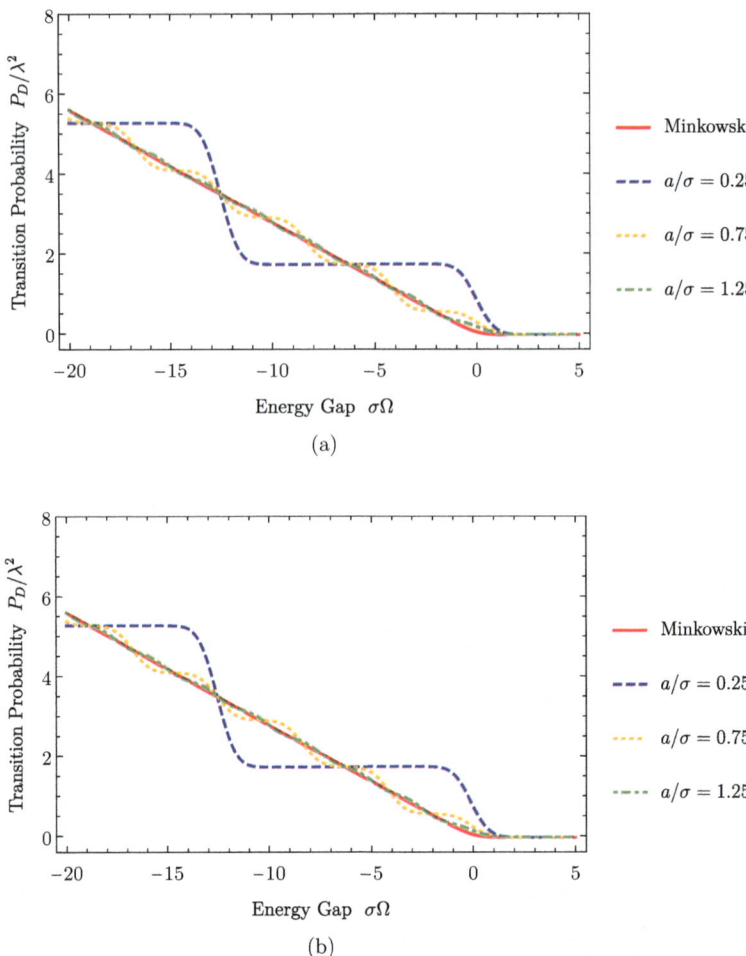

Fig. 4.3 The transition probability of a detector coupled to an (**a**) untwisted ($\eta = 1$) and (**b**) twisted ($\eta = -1$) massless scalar field in Minkowski space $\mathcal{M}$ (solid red) is compared to the transition probability of the same detector in the cylindrical spacetime $\mathcal{M}_-$ (broken lines) by plotting the transition probability as a function of the energy gap $\sigma \Omega$ of the detector. The detector is located a distance $\sqrt{d_x^2 + d_y^2}/\sigma = 5$ away from the origin in the xy-plane. Different circumferences a/σ of the cylindrical spacetime are shown

4.3   Entanglement and Correlation Harvesting in Quotients of Minkowski Space

In this section we apply the entanglement harvesting protocol developed in Chap. 3 to Minkowski space $\mathcal{M}$ and the two cylindrical spacetimes $\mathcal{M}_0$ and $\mathcal{M}_-$ introduced in Sect. 4.1. We begin by computing the matrix elements X and C, defined in

Eqs. (3.26) and (3.25), appearing in the final state $\rho_{AB} \in \mathcal{S}(\mathcal{H}_A \otimes \mathcal{H}_B)$ of two Unruh-DeWitt detectors in all three spacetimes $\mathcal{M}$, $\mathcal{M}_0$, and $\mathcal{M}_-$. We then use these results to compute the concurrence associated with ρ_{AB}, quantifying how entangled the detectors have become as a result of their interaction with the field, and the correlation between local measurements of the Pauli z operator on the final state of each detector.

In all three spacetimes we will consider the trajectories of detector A and B to be

$$x_A(\tau) := \{t = \tau_A, \, \mathbf{x}_A = (x_A, y_A, z_A)\}, \tag{4.38a}$$

$$x_B(\tau) := \{t = \tau_B, \, \mathbf{x}_B = (x_B, y_B, z_B)\}, \tag{4.38b}$$

where τ_A and τ_B are the proper time of each detector. The detectors are at rest with respect to one another and with respect to the chosen coordinate frame. This allows us to parametrize both their trajectories with the coordinate time t.

We will again suppose that the switching function of each detector is Gaussian

$$\chi_A(t) = \chi_B(t) = e^{-t^2/2\sigma^2}, \tag{4.39}$$

with the interpretation that each detector is interacting with the field for an approximate amount of proper time $k\sigma$. We will also suppose that the energy gap of each detector is the same, $\Omega_A = \Omega_B = \Omega$.

4.3.1 Computation of the Matrix Element X

With the above choices for the detectors' switching functions and trajectories, the matrix element X defined in Eq. (3.26) simplifies to

$$X = -\lambda^2 \int_{-\infty}^{\infty} dt \int_{-\infty}^{t} dt' \, e^{-(t'+t)^2/4\sigma^2} e^{-(t'-t)^2/4\sigma^2} e^{-i\Omega(t+t')}$$

$$\times \left[W\big(x_A(t'), x_B(t)\big) + W\big(x_B(t'), x_A(t)\big) \right]. \tag{4.40}$$

Using Eq. (4.40) we now compute X in Minkowski space $\mathcal{M}$ and the two cylindrical spacetimes $\mathcal{M}_0$ and $\mathcal{M}_-$. The reader uninterested in the explicit evaluation of X may skip to Table 4.2 where the values of X are summarized in all three spacetimes.

Evaluation of X in Minkowski Space $\mathcal{M}$

We begin by evaluating X in Minkowski space $\mathcal{M}$. Using the Minkowski space Wightman function given in Eq. (4.10) and the detectors' trajectories in Eq. (4.38), the factor in the square brackets in Eq. (4.40) evaluates to

$$2\left(\frac{1}{4\pi i}\,\mathrm{sgn}(t'-t)\delta\big[(t'-t)^2-L^2\big]-\frac{1}{4\pi^2\big[(t'-t)^2-L^2\big]}\right),\qquad(4.41)$$

where $L^2 := (x_A - x_B)^2 + (y_A - y_B)^2 + (z_A - z_B)^2$ is the square of the spatial distance between the two detectors.

Upon substituting Eq. (4.41) into Eq. (4.40) and changing the integration variables to $u' = t' + t$ and $u = t' - t$, the quantity X simplifies to

$$X_{\mathcal{M}} = -\lambda^2 \int_{-\infty}^{\infty} du'\, e^{-u'^2/4\sigma^2} e^{-i\Omega u'} \int_{-\infty}^{0} du\, e^{-u^2/4\sigma^2}$$

$$\times \left(\frac{1}{4\pi i}\,\mathrm{sgn}(u)\delta\big[u^2-L^2\big]-\frac{1}{4\pi^2\big[u^2-L^2\big]}\right)$$

$$= -\lambda^2 \int_{-\infty}^{\infty} du'\, e^{-u'^2/4\sigma^2} e^{-i\Omega u'} \int_{0}^{\infty} du\, e^{-u^2/4\sigma^2}$$

$$\times \left(\frac{1}{4\pi i}\,\mathrm{sgn}(-u)\delta\big[u^2-L^2\big]-\frac{1}{4\pi^2\big[u^2-L^2\big]}\right)$$

$$= 2\sqrt{\pi}\lambda^2\sigma e^{-\sigma^2\Omega^2} \int_{0}^{\infty} du\, e^{-u^2/4\sigma^2}\left(\frac{1}{4\pi i}\,\mathrm{sgn}(u)\delta\big[u^2-L^2\big]+\frac{1}{4\pi^2\big[u^2-L^2\big]}\right)$$

$$= 2\sqrt{\pi}\lambda^2\sigma e^{-\sigma^2\Omega^2} \int_{0}^{\infty} du\, e^{-u^2/4\sigma^2}\left(\frac{1}{4\pi i}\frac{1}{2L}\delta[u-L]+\frac{1}{4\pi^2\big[u^2-L^2\big]}\right)$$

$$= 2\sqrt{\pi}\lambda^2\sigma e^{-\sigma^2\Omega^2}\left(\frac{1}{4\pi i}\frac{e^{-L^2/4\sigma^2}}{2L}+\frac{1}{4\pi^2}\left[i\frac{\pi}{2L}e^{-L^2/4\sigma^2}\,\mathrm{erf}\!\left(i\frac{L}{2\sigma}\right)\right]\right)$$

$$= i\frac{\lambda^2}{4\sqrt{\pi}}\frac{\sigma}{L}e^{-\sigma^2\Omega^2-L^2/4\sigma^2}\left[\mathrm{erf}\!\left(i\frac{L}{2\sigma}\right)-1\right],\qquad(4.42)$$

where the integration was performed with Mathematica.

Evaluation of X in the Cylindrical Spacetime $\mathcal{M}_0$

Again, let us begin by evaluating the terms inside the square brackets in Eq. (4.40) using the Wightman function in $\mathcal{M}_0$ given in Eq. (4.12)

$$2\sum_{n=-\infty}^{\infty}\eta^n\left(\frac{1}{4\pi i}\,\mathrm{sgn}(t'-t)\delta\big[(t'-t)^2-L_{\mathcal{M}_0}(n)^2\big]\right.$$

$$\left.-\frac{1}{4\pi^2\big[(t'-t)^2-L_{\mathcal{M}_0}(n)^2\big]}\right),\qquad(4.43)$$

where

$$L_{\mathcal{M}_0}(n)^2 := L^2 + a^2n^2 + 2anL\sin\theta,\qquad(4.44)$$

with θ being the angle between a line connecting the two detectors and the z-axis.

Upon comparison with Eq. (4.41), it is clear that each term in the sum in Eq. (4.43) is identical to Eq. (4.41) with the replacement $L \to L_{\mathcal{M}_0}(n)$. Therefore, $X_{\mathcal{M}_0}$ is given by Eq. (4.48) with the replacement $L \to L_{\mathcal{M}_0}(n)$ and a summation over n

$$
X_{\mathcal{M}_0} = X_{\mathcal{M}} + i \frac{\lambda^2}{4\sqrt{\pi}} \sum_{n\neq 0} \eta^n \frac{\sigma}{L_{\mathcal{M}_0}(n)} e^{-\sigma^2 \Omega^2 - L_{\mathcal{M}_0}(n)^2/4\sigma_D^2}
$$
$$
\times \left[\text{erf}\left(i\frac{L_{\mathcal{M}_0}(n)^2}{2\sigma} \right) - 1 \right].
\tag{4.45}
$$

Evaluation of X in the Cylindrical Spacetime $\mathcal{M}_-$

Evaluating the terms inside the square brackets in Eq. (4.40), using the Wightman function in $\mathcal{M}_-$ given in Eq. (4.14), yields

$$
2 \sum_{n=-\infty}^{\infty} \eta^n \left(\frac{1}{4\pi i} \text{sgn}(t' - t)\delta\left[(t' - t)^2 - L_{\mathcal{M}_-}(n)^2 \right] \right.
$$
$$
\left. - \frac{1}{4\pi^2 \left[(t' - t)^2 - L_{\mathcal{M}_-}(n)^2 \right]} \right),
\tag{4.46}
$$

where

$$
L_{\mathcal{M}_-}(n)^2 := L_{\mathcal{M}_0}(n)^2 + 2\left[1 - (-1)^n \right] \mathbf{d}_A \cdot \mathbf{d}_B,
\tag{4.47}
$$

and $\mathbf{d}_A := (x_A, y_A)$ and $\mathbf{d}_B := (x_B, y_B)$ are vectors lying in the xy-plane.

Again, upon comparison with Eq. (4.41) it is clear that each term in the sum in Eq. (4.46) is identical to Eq. (4.41) with the replacement $L \to L_{\mathcal{M}_-}(n)$. Therefore, $X_{\mathcal{M}_-}$ is given by Eq. (4.48) with the replacement $L \to L_{\mathcal{M}_-}(n)$ and a summation over n

$$
X_{\mathcal{M}_-} = X_{\mathcal{M}} + i \frac{\lambda^2}{4\sqrt{\pi}} \sum_{n\neq 0} \eta^n \frac{\sigma}{L_{\mathcal{M}_-}(n)} e^{-\sigma^2 \Omega^2 - L_{\mathcal{M}_-}(n)^2/4\sigma^2}
$$
$$
\times \left[\text{erf}\left(i\frac{L_{\mathcal{M}_-}(n)}{2\sigma} \right) - 1 \right].
\tag{4.48}
$$

Summary

The above evaluations of the matrix element X in all three spacetimes $\mathcal{M}$, $\mathcal{M}_0$, and $\mathcal{M}_-$ are summarized in Table 4.2.

Table 4.2 The matrix element X in Minkowski space $\mathcal{M}$ and the two cylindrical spacetimes $\mathcal{M}_0$ and $\mathcal{M}_-$ constructed from the identifications $(t, x, y, z) \sim (t, x, y, z + a)$ and $(t, x, y, z) \sim (t, -x, -y, z + a)$ of Minkowski space, respectively

Spacetime	The matrix element X to leading order in λ
$\mathcal{M}$	$X_{\mathcal{M}} = i \dfrac{\lambda^2}{4\sqrt{\pi}} \dfrac{\sigma}{L} e^{-\sigma^2 \Omega^2 - L^2/4\sigma^2} \left[\mathrm{erf}\left(i \dfrac{L}{2\sigma}\right) - 1 \right]$
$\mathcal{M}_0$	$X_{\mathcal{M}_0} = X_{\mathcal{M}}$ $+ i \dfrac{\lambda^2}{4\sqrt{\pi}} \sum_{n \neq 0} \eta^n \dfrac{\sigma}{L_{\mathcal{M}_0}(n)} e^{-\sigma^2 \Omega^2 - L_{\mathcal{M}_0}(n)^2/4\sigma_D^2} \left[\mathrm{erf}\left(i \dfrac{L_{\mathcal{M}_0}(n)^2}{2\sigma}\right) - 1 \right]$ where $L_{\mathcal{M}_0}(n)^2 := L^2 + a^2 n^2 + 2an L \sin\theta$
$\mathcal{M}_-$	$X_{\mathcal{M}_-} = X_{\mathcal{M}}$ $+ i \dfrac{\lambda^2}{4\sqrt{\pi}} \sum_{n \neq 0} \eta^n \dfrac{\sigma}{L_{\mathcal{M}_-}(n)} e^{-\sigma^2 \Omega^2 - L_{\mathcal{M}_-}(n)^2/4\sigma^2} \left[\mathrm{erf}\left(i \dfrac{L_{\mathcal{M}_-}(n)}{2\sigma}\right) - 1 \right]$ where $L_{\mathcal{M}_-}(n)^2 := L_{\mathcal{M}_0}(n)^2 + 2\left[1 - (-1)^n\right] \mathbf{d}_A \cdot \mathbf{d}_B$

The trajectory of the two detectors is $x_A(t) = (t, x_A, y_A, z_A)$ and $x_B(t) = (t, x_B, y_B, z_B)$, and $\mathbf{d}_A := (x_A, y_A)$, $\mathbf{d}_B := (x_B, y_B)$, L is the spatial distance between the two detectors, and $\sin\theta := |z_A - z_B|/L$

4.3.2 Computation of the Matrix Element C

With the above choices for the detectors' switching functions and trajectories, the matrix element C defined in Eq. (3.25) simplifies to

$$C = \lambda^2 \int dt dt' \, e^{-(t'+t)^2/4\sigma^2} e^{-(t'-t)^2/4\sigma^2} e^{-i\Omega(t'-t)} W\big(x_A(t'), x_B(t)\big). \quad (4.49)$$

Using Eq. (4.49), we now compute C in Minkowski space $\mathcal{M}$ and both cylindrical spacetimes $\mathcal{M}_0$ and $\mathcal{M}_-$. The reader uninterested in the explicit evaluation of Eq. (4.49) may skip to Table 4.3 where C in all three spacetimes is summarized.

Evaluation of C in Minkowski Space $\mathcal{M}$

The Wightman function in Minkowski space evaluated along the detectors' trajectories appearing in Eq. (4.49) yields

$$W_{\mathcal{M}}\big(x_A(t'), x_B(t)\big) = \frac{1}{4\pi i} \, \mathrm{sgn}\left(t' - t\right) \delta\left((t' - t)^2 - L^2\right)$$
$$- \frac{1}{4\pi^2\left((t' - t)^2 - L^2\right)}. \quad (4.50)$$

Upon substituting Eq. (4.50) into Eq. (4.49) and changing the integration variables to $u = t' - t$ and $u' = t' + t$, the matrix element C simplifies to

$$C_{\mathcal{M}} = \lambda^2 \frac{1}{2} \int du' du \, e^{-u'^2/4\sigma} e^{-u^2/4\sigma} e^{-i\Omega u}$$

$$\times \left[\frac{1}{4\pi i} \, \mathrm{sgn}\,(u) \, \delta\left(u^2 - L^2\right) - \frac{1}{4\pi^2\left(u^2 - L^2\right)} \right]$$

$$= \lambda^2 \sqrt{\pi}\sigma \int du \, e^{-\frac{u^2}{4\sigma^2}} e^{-i\Omega u}$$

$$\times \left[\frac{1}{4\pi i} \frac{\mathrm{sgn}\,(u)}{2|L|} \left[\delta\left(u + L\right) + \left(u - L\right)\right] - \frac{1}{4\pi^2\left(u^2 - L^2\right)} \right]$$

$$= -\lambda^2 \frac{\sigma}{4\sqrt{\pi}} \left[\frac{e^{-\frac{L^2}{4\sigma^2}}}{L} \sin\left(\Omega L\right) + \frac{1}{\pi} \int du \, \frac{e^{-\frac{u^2}{4\sigma^2}} e^{-i\Omega u}}{u^2 - L^2} \right]. \qquad (4.51)$$

The remaining integral in Eq. (4.51) may be evaluated using the convolution theorem

$$\int du \, \frac{e^{-\frac{u'^2}{4\sigma^2}} e^{-i\Omega u'}}{u^2 - L^2} = \int dudu' \, \delta(u' - u) \frac{e^{-u^2/4\sigma^2} e^{-i\Omega u}}{u^2 - L^2}$$

$$= \int dudu' \left(\frac{1}{2\pi} \int dy \, e^{iy(u'-u)} \right) \frac{e^{-u'^2/4\sigma^2} e^{-i\Omega u'}}{u^2 - L^2}$$

$$= \frac{1}{2\pi} \int dy \left(\int du' \, e^{-u'^2/4\sigma^2} e^{-iu'(\Omega - y)} \right) \left(\int du \, \frac{e^{-iyu}}{u^2 - L^2} \right)$$

$$= \frac{1}{2\pi} \int dy \left(2\sqrt{\pi}\sigma e^{-\sigma^2(\Omega - y)^2} \right) \left(-\pi \frac{\mathrm{sgn}(y)}{L} \sin Ly \right)$$

$$= -\frac{\pi}{L} e^{-L^2/4\sigma^2} \mathrm{Im}\left[e^{iL\Omega} \, \mathrm{erf}\left(i\frac{L}{2\sigma} + \sigma\Omega \right) \right]. \qquad (4.52)$$

After substituting Eq. (4.52) into Eq. (4.51), we find C to be

$$C_{\mathcal{M}} = \frac{\lambda^2}{4\sqrt{\pi}} \frac{\sigma}{L} e^{-L^2/4\sigma^2} \left(\mathrm{Im}\left[e^{iL\Omega} \, \mathrm{erf}\left(i\frac{L}{2\sigma} + \sigma\Omega \right) \right] - \sin \Omega L \right). \qquad (4.53)$$

Evaluation of C in the Cylindrical Spacetime $\mathcal{M}_0$

The Wightman function in the cylindrical spacetime $\mathcal{M}_0$ evaluated along the detectors' trajectories appearing in Eq. (4.49) is

$$W_{\mathcal{M}_0}\big(x_A(t'), x_B(t)\big) = \sum_{n=-\infty}^{\infty} \eta^n \left(\frac{1}{4\pi i} \operatorname{sgn}(t' - t)\delta\big[(t' - t)^2 - L_{\mathcal{M}_0}(n)^2\big] \right.$$

$$\left. - \frac{1}{4\pi^2 \big[(t' - t)^2 - L_{\mathcal{M}_0}(n)^2\big]} \right). \tag{4.54}$$

Upon comparison with Eq. (4.50), it is clear that each term in the sum in Eq. (4.54) is identical to Eq. (4.50) with the replacement $L \to L_{\mathcal{M}_0}(n)$. Therefore, $C_{\mathcal{M}_0}$ is given by Eq. (4.53) with the replacement $L \to L_{\mathcal{M}_0}(n)$ and a summation over n

$$C_{\mathcal{M}_0} = C_{\mathcal{M}} + \frac{\lambda^2}{4\sqrt{\pi}} \sum_{n \neq 0} \eta^n \frac{\sigma}{L_{\mathcal{M}_0}(n)} e^{-L_{\mathcal{M}_0}(n)^2/4\sigma^2}$$

$$\times \left(\operatorname{Im}\left[e^{i L_{\mathcal{M}_0}(n)\Omega} \operatorname{erf}\left(i\frac{L_{\mathcal{M}_0}(n)}{2\sigma} + \sigma\Omega \right) \right] - \sin \Omega L_{\mathcal{M}_0}(n) \right). \tag{4.55}$$

Evaluation of C in the Cylindrical Spacetime $\mathcal{M}_-$

The Wightman function in the cylindrical spacetime $\mathcal{M}_-$ evaluated along the detectors' trajectories appearing in Eq. (4.49) is

$$W_{\mathcal{M}_-}\big(x_A(t'), x_B(t)\big) = \sum_{n=-\infty}^{\infty} \eta^n \left(\frac{1}{4\pi i} \operatorname{sgn}(t' - t)\delta\big[(t' - t)^2 - L_{\mathcal{M}_-}(n)^2\big] \right.$$

$$\left. - \frac{1}{4\pi^2 \big[(t' - t)^2 - L_{\mathcal{M}_-}(n)^2\big]} \right). \tag{4.56}$$

Again, upon comparison with Eq. (4.50), it is clear that each term in the sum in Eq. (4.56) is identical to Eq. (4.50) with the replacement $L \to L_{\mathcal{M}_-}(n)$. Therefore, $C_{\mathcal{M}_-}$ is given by Eq. (4.53) with the replacement $L \to L_{\mathcal{M}_-}(n)$ and a summation over n

$$C_{\mathcal{M}_-} = C_{\mathcal{M}} + \frac{\lambda^2}{4\sqrt{\pi}} \sum_{n \neq 0} \eta^n \frac{\sigma}{L_{\mathcal{M}_-}(n)} e^{-L_{\mathcal{M}_-}(n)^2/4\sigma^2}$$

$$\times \left(\operatorname{Im}\left[e^{i L_{\mathcal{M}_-}(n)\Omega} \operatorname{erf}\left(i\frac{L_{\mathcal{M}_-}(n)}{2\sigma} + \sigma\Omega \right) \right] - \sin \Omega L_{\mathcal{M}_-}(n) \right).$$

$$\tag{4.57}$$

Summary

The above evaluations of the matrix element C in all three spacetimes $\mathcal{M}$, $\mathcal{M}_0$, and $\mathcal{M}_-$ are summarized in Table 4.3.

4.3.3 Harvesting Entanglement and Correlations in $\mathcal{M}$, $\mathcal{M}_0$, and $\mathcal{M}_-$

Having evaluated the matrix elements X and C, and the transition probability P_D in Sect. 4.2, for detectors in $\mathcal{M}$, $\mathcal{M}_0$, and $\mathcal{M}_-$, we now compare how entangled the final state ρ_{AB} of two static detectors is in these spacetimes. We quantify this entanglement with the concurrence $\mathcal{C}(\rho_{AB})$, which is plotted in Figs. 4.4, 4.5, 4.6, 4.7, 4.8, 4.9, and 4.10. In addition, we examine the correlations between local measurements of the Pauli z operator σ_z by evaluating the correlation function $\operatorname{corr}(\rho_{AB})$ in all three spacetimes.

We first note from Tables 4.2 and 4.3 that the matrix elements X and C both diverge when the separation of the detectors goes to zero, consequently so does the entanglement and correlations between the detectors. This divergence is due to the

Table 4.3 The matrix element C in Minkowski space $\mathcal{M}$ and the two cylindrical universes $\mathcal{M}_0$ and $\mathcal{M}_-$ constructed from the identifications $(t, x, y, z) \sim (t, x, y, z + a)$ and $(t, x, y, z) \sim (t, -x, -y, z + a)$ of Minkowski space, respectively

Spacetime	The matrix element C to leading order in λ
$\mathcal{M}$	$C_\mathcal{M} = \dfrac{\lambda^2}{4\sqrt{\pi}} \dfrac{\sigma}{L} e^{-L^2/4\sigma^2} \left(\operatorname{Im}\left[e^{iL\Omega} \operatorname{erf}\left(i\dfrac{L}{2\sigma} + \sigma\Omega \right) \right] - \sin \Omega L \right)$
$\mathcal{M}_0$	$C_{\mathcal{M}_0} = C_\mathcal{M} + \dfrac{\lambda^2}{4\sqrt{\pi}} \sum_{n\neq 0} \eta^n \dfrac{\sigma}{L_{\mathcal{M}_0}(n)} e^{-L_{\mathcal{M}_0}(n)^2/4\sigma^2}$ $\times \left(\operatorname{Im}\left[e^{iL_{\mathcal{M}_0}(n)\Omega} \operatorname{erf}\left(i\dfrac{L_{\mathcal{M}_0}(n)}{2\sigma} + \sigma\Omega \right) \right] - \sin \Omega L_{\mathcal{M}_0}(n) \right)$ where $L_{\mathcal{M}_0}(n)^2 := L^2 + a^2 n^2 + 2anL \sin \theta$
$\mathcal{M}_-$	$C_{\mathcal{M}_-} = C_\mathcal{M} + \dfrac{\lambda^2}{4\sqrt{\pi}} \sum_{n\neq 0} \eta^n \dfrac{\sigma}{L_{\mathcal{M}_-}(n)} e^{-L_{\mathcal{M}_-}(n)^2/4\sigma^2}$ $\times \left(\operatorname{Im}\left[e^{iL_{\mathcal{M}_-}(n)\Omega} \operatorname{erf}\left(i\dfrac{L_{\mathcal{M}_-}(n)}{2\sigma} + \sigma\Omega \right) \right] - \sin \Omega L_{\mathcal{M}_-}(n) \right)$ where $L_{\mathcal{M}_-}(n)^2 := L_{\mathcal{M}_0}(n)^2 + 2\left[1 - (-1)^n \right] \mathbf{d}_A \cdot \mathbf{d}_B$

The trajectory of the two detectors is $x_A(t) = (t, x_A, y_A, z_A)$ and $x_B(t) = (t, x_B, y_B, z_B)$, and $\mathbf{d}_A := (x_A, y_A)$, $\mathbf{d}_B := (x_B, y_B)$, L is the spatial distance between the two detectors, and $\sin \theta := |z_A - z_B|/L$

fact that the detector couples to the field at a point, and could have been regulated had the detector been coupled to a smeared field modelling the spatial extent of the detector [5].

In Figs. 4.5 and 4.8 the contours indicate the value of the concurrence in Minkowski space, and in Figs. 4.6 and 4.9 the contours indicate the value of the correlation function in Minkowski space. In Figs. 4.5, 4.6, 4.8, and 4.9, the thick black line denotes the boundary to the left of which the entanglement in the final state of the detectors is identically zero. For plotting purposes, the image sums appearing in the transition probabilities P_A and P_B and the matrix elements X and C have been truncated after 100 terms, which results in an error on the order of 10^{-5}.

Description of Fig. 4.4

Figure 4.4 depicts both the concurrence $\mathcal{C}(\rho_{AB})_{\mathcal{M}}$ and correlation function $\text{corr}(\rho_{AB})_{\mathcal{M}}$ for two detectors in Minkowski space $\mathcal{M}$ as a function of the detector separation L/σ and energy gap $\sigma\Omega$ in units of the interaction length σ.

From Fig. 4.4 we see that the detectors become most entangled and strongly correlated when they have a small positive energy gap. Furthermore, as the detector separation increases, both the entanglement and correlations decrease. This should have been expected from the fact that the Wightman function $W_{\mathcal{M}}(x, x')$ decreases as the distance between the spacetime points x and x' increases.

We also observe that the entanglement in the final state of the detectors vanishes for the region to the left of the thick black line, while the correlations decay to zero smoothly.

Description of Fig. 4.5

In Fig. 4.5 the concurrence of the final state of two detectors in the cylindrical spacetime $\mathcal{M}_0$ is compared to the same quantity in Minkowski space $\mathcal{M}$ by plotting their difference $\mathcal{C}(\rho_{AB})_{\mathcal{M}_0} - \mathcal{C}(\rho_{AB})_{\mathcal{M}}$. The circumference of the spacetime $\mathcal{M}_0$ is $a/\sigma = 4$ and the detectors are aligned with the identified direction.

From Fig. 4.5 we observe that for detectors coupled to untwisted or twisted fields, the region in which ρ_{AB} is not entangled coincides closely with the region in which the detectors are not entangled in Minkowski space, i.e. the region to the left of the thick black line.

We also observe that for energy gaps where the detectors are most entangled, detectors coupled to an untwisted (a twisted) field are less (more) entangled than identical detectors in Minkowski space. This difference between detectors coupled to twisted and untwisted fields is in large part due to the difference in the transition probability P_D of a single detector. Since the concurrence is given by $\mathcal{C}(\rho_{AB}) = 2 \max\left[0, |X| - \sqrt{P_A P_B}\right] + \mathcal{O}(\lambda^4)$, for fixed $|X|$, we see that the smaller the transition probabilities P_A and P_B, the greater the concurrence quantifying the entanglement between the detectors. From Fig. 4.1, we see that the transition probability of

Fig. 4.4 For two Unruh-DeWitt detectors in Minkowski space $\mathcal{M}$, (**a**) the concurrence $\mathcal{C}(\rho_{AB})_\mathcal{M}$ and (**b**) correlation function corr$(\rho_{AB})_\mathcal{M}$ are plotted as a function of their separation L/σ and energy gap $\sigma\Omega$. In both (**a**) and (**b**), to the left of the thick black line the final state of the two detectors ρ_{AB} is not entangled; this is also true for Figs. 4.5, 4.6, 4.8, and 4.9

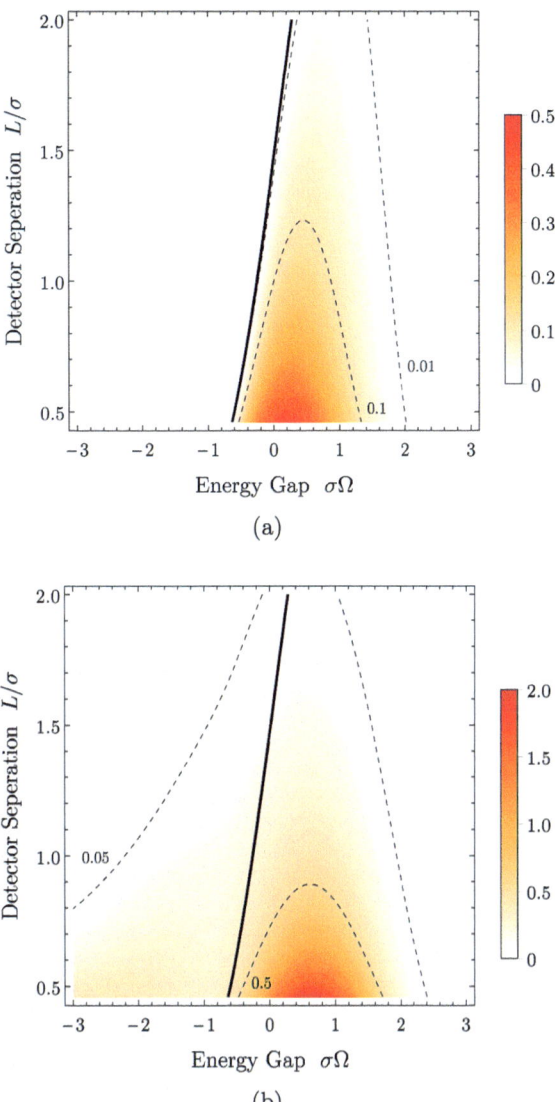

a detector in $\mathcal{M}_0$ coupled to a twisted field is much less than a detector coupled to an untwisted field for detectors with a small positive or negative energy gap.

However, for detectors with a large positive energy gap, we see that the entanglement between the detectors is greatest for detectors coupled to an untwisted field. This is the regime in which the transition probability of a detector coupled to either an untwisted or twisted field is comparable.

Fig. 4.5 The difference between the concurrence of the final state of two detectors in the cylindrical spacetime $\mathcal{M}_0$ and Minkowski space $\mathcal{M}$, $C(\rho_{AB})_{\mathcal{M}_0} - C(\rho_{AB})_{\mathcal{M}}$, is plotted as a function of their separation L/σ and energy gap $\sigma\Omega$ for detectors coupled to (**a**) untwisted fields ($\eta = 1$) and (**b**) twisted fields ($\eta = -1$). In both (**a**) and (**b**) the circumference of the spacetime is $a/\sigma = 4$ and the detectors are aligned with the identified direction, $\theta = 0$

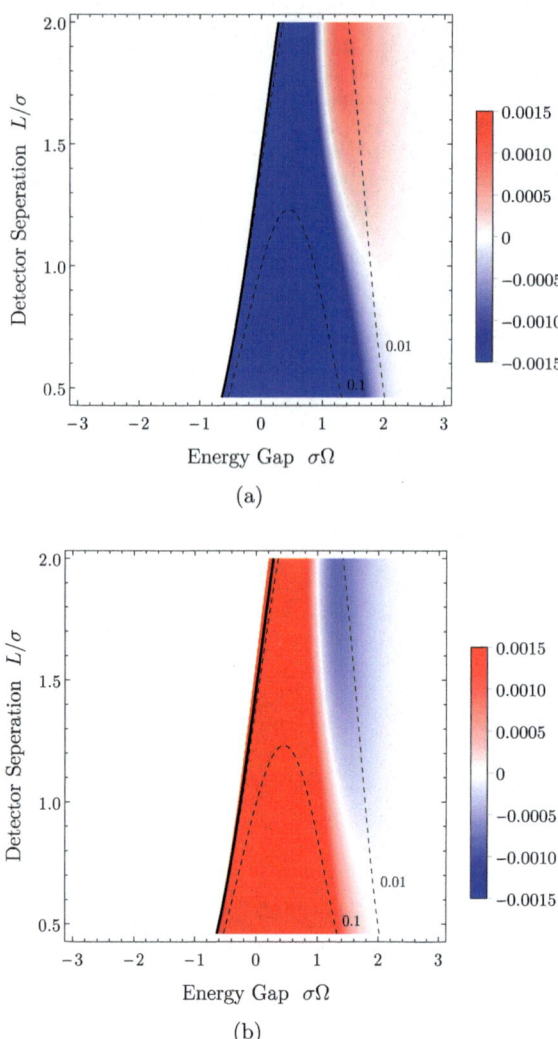

Description of Fig. 4.6

In Fig. 4.6 we examine the correlation between the outcomes of local measurements of the Pauli z operator on each detector, given the detectors are in their final state ρ_{AB}. This is done for detectors in the cylindrical spacetime $\mathcal{M}_0$ and compared to detectors in Minkowski space $\mathcal{M}$ by plotting the difference in the correlation functions $\mathrm{corr}(\rho_{AB})_{\mathcal{M}_0} - \mathrm{corr}(\rho_{AB})_{\mathcal{M}}$. The circumference of the spacetime $\mathcal{M}_0$ is $a/\sigma = 4$ and the detectors are aligned with the identified direction.

For detectors with a positive energy gap, the correlations in the final state of the two detectors in $\mathcal{M}_0$ as compared to equivalent detectors in $\mathcal{M}$ behave as the

Fig. 4.6 The difference between the correlation function associated with the final state of two detectors in the cylindrical spacetime $\mathcal{M}_0$ and Minkowski space $\mathcal{M}$, $\mathrm{corr}(\rho_{AB})_{\mathcal{M}_0} - \mathrm{corr}(\rho_{AB})_{\mathcal{M}}$, is plotted as a function of their separation L/σ and energy gap $\sigma\Omega$ for detectors coupled to (**a**) untwisted fields ($\eta = 1$) and (**b**) twisted fields ($\eta = -1$). In both (**a**) and (**b**) the circumference of the spacetime is $a/\sigma = 4$ and the detectors are aligned with the identified direction, $\theta = 0$

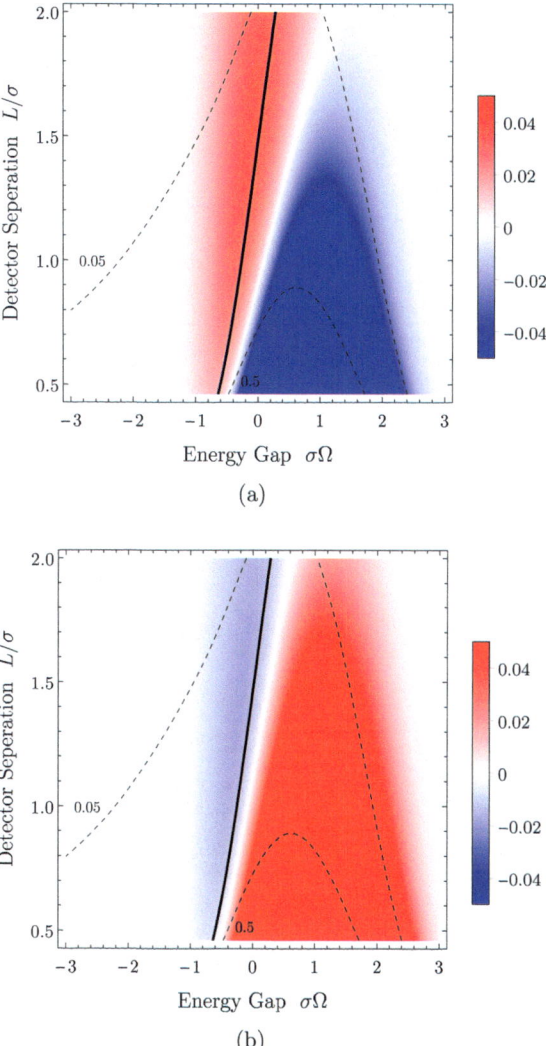

entanglement depicted in Fig. 4.5. However, for negative energy gaps immediately to the right of the thick black line, in the region where the entanglement between the two detectors vanishes, we see that the correlations are greater (smaller) for detectors coupled to an untwisted (twisted) field as compared to correlations between detectors in Minkowski space. The converse is true when the energy gap becomes more negative ($\sigma\Omega \approx -3$).

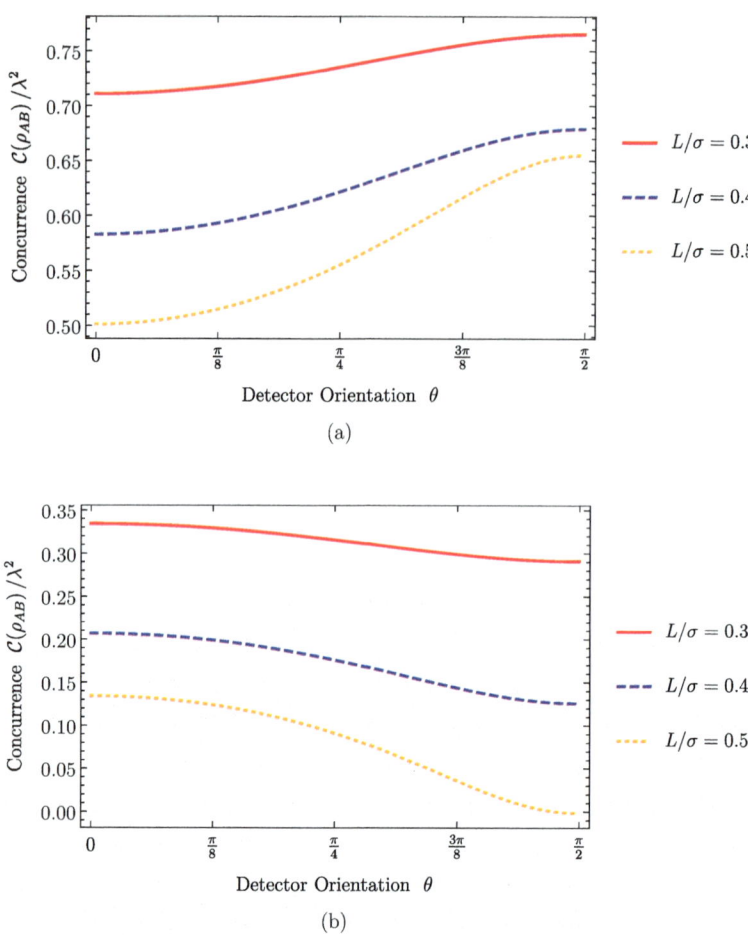

Fig. 4.7 For two detectors in the cylindrical spacetime $\mathcal{M}_0$, the concurrence associated with the final state of the two detectors $\mathcal{C}(\rho_{AB})_{\mathcal{M}_0}$ is plotted as a function of the detectors orientation with respect to the identified direction. When $\theta = 0$ ($\theta = \pi/2$) the detectors are aligned with (orthogonal to) the identified direction. The energy gap of the detectors is $\sigma\Omega = 0.75$. This is done for detectors coupled to (**a**) untwisted fields ($\eta = 1$) and (**b**) twisted fields ($\eta = -1$)

Description of Fig. 4.7

In Fig. 4.7 the concurrence $\mathcal{C}(\rho_{AB})_{\mathcal{M}_0}$ of the final state of two detectors in the cylindrical spacetime $\mathcal{M}_0$ is plotted as a function of the detectors orientation with respect to the identified direction (i.e. the z axis); $\theta = 0$ ($\theta = \pi/2$) the detectors are aligned with (orthogonal to) the identified direction. The energy gap of the detector is $\sigma\Omega = 0.75$.

We first observe that for detectors coupled to either twisted or untwisted fields with the chosen energy gap, the amount of entanglement that results in the final state of the two detectors depends on their orientation with respect to the identified direction. For large detector separation, the total amount of entanglement in the final state is less than for small detector separation; however, the dependence on the detector orientation is greater. For untwisted (twisted) fields and detectors with the energy gap plotted in Fig. 4.7, the entanglement in the final state increases (decreases) as the angle with respect to the identified direction increases.

Description of Fig. 4.8

In Fig. 4.8 the concurrence of the final state of two detectors in the cylindrical spacetime $\mathcal{M}_-$ is compared to the same quantity in Minkowski space $\mathcal{M}$ by plotting their difference $\mathcal{C}(\rho_{AB})_{\mathcal{M}_-} - \mathcal{C}(\rho_{AB})_{\mathcal{M}}$. The circumference of the spacetime $\mathcal{M}_-$ is $a/\sigma = 4$, and $\mathbf{d}_A = \mathbf{d}_B$, $|\mathbf{d}_A|/\sigma = 0.25$, and the detectors are aligned with the identified direction.

From Fig. 4.8, we observe that for detectors coupled to either untwisted or twisted fields, the region in which ρ_{AB} is not entangled is approximately the same region in which detectors are not entangled in Minkowski space, i.e. the region to the left of the thick black line. A second observation is that the concurrence associated with the final state of the two detectors in $\mathcal{M}_-$ is greater (less) than the same quantity in Minkowski space $\mathcal{M}$ for twisted (untwisted) fields. This is similar to detectors in $\mathcal{M}_0$, as shown in Fig. 4.5.

Description of Fig. 4.9

In Fig. 4.9 we examine the correlation between the outcomes of local measurements of the Pauli z operator on each detector, given the detectors are in their final state ρ_{AB}. This is done for detectors in the cylindrical spacetime $\mathcal{M}_-$ and compared to detectors in Minkowski space $\mathcal{M}$ by plotting the difference in the correlation functions $\mathrm{corr}(\rho_{AB})_{\mathcal{M}_-} - \mathrm{corr}(\rho_{AB})_{\mathcal{M}}$. The circumference of the spacetime $\mathcal{M}_-$ is $a/\sigma = 4$, and $\mathbf{d}_A = \mathbf{d}_B$, $|\mathbf{d}_A|/\sigma = 0.25$, and the detectors are aligned with the identified direction.

Upon comparison of Fig. 4.9 with Fig. 4.6, we observe that the correlation function associated with detectors coupled to either untwisted or twisted fields in the cylindrical spacetime $\mathcal{M}_-$ behaves similar to the correlation function associated with detectors in $\mathcal{M}_0$; see the discussion of Fig. 4.6.

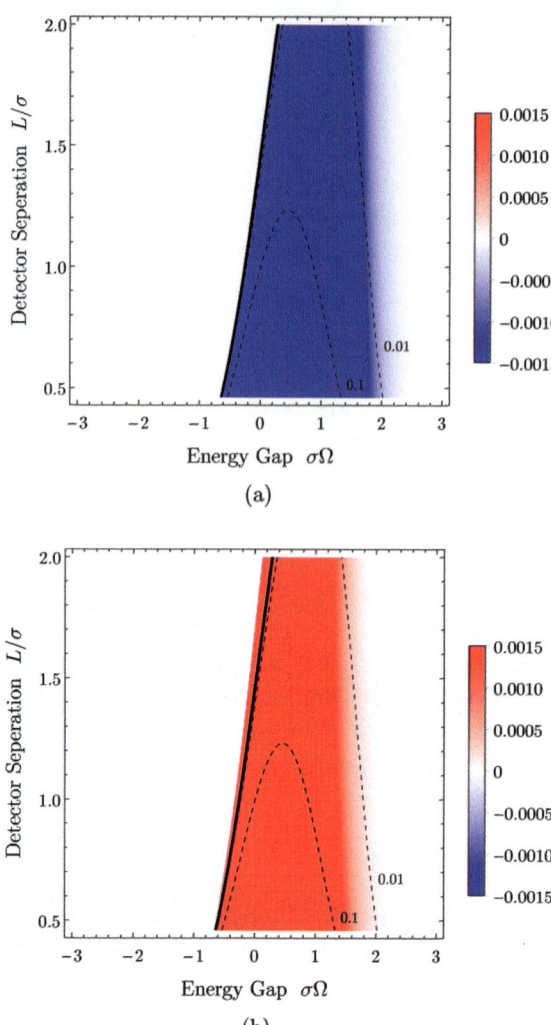

Fig. 4.8 The difference between the concurrence of the final state of two detectors in the cylindrical spacetime $\mathcal{M}_-$ and Minkowski space $\mathcal{M}$, $\mathcal{C}(\rho_{AB})_{\mathcal{M}_-} - \mathcal{C}(\rho_{AB})_{\mathcal{M}}$, is plotted as a function of their separation L/σ and energy gap $\sigma\Omega$, for detectors coupled to (**a**) untwisted fields ($\eta = 1$) and (**b**) twisted fields ($\eta = -1$). In both (**a**) and (**b**) the circumference of the spacetime is $a/\sigma = 4$, $\mathbf{d}_A = \mathbf{d}_B$, $|\mathbf{d}_A|/\sigma = 0.25$, and $\theta = 0$

Description of Fig. 4.10

In Fig. 4.10 the concurrence $\mathcal{C}(\rho_{AB})_{\mathcal{M}_-}$ of the final state of two detectors in the cylindrical spacetime $\mathcal{M}_-$ is plotted as a function of the detectors orientation with respect to the identified direction (the z direction); $\theta = 0$ ($\theta = \pi/2$) the detectors are aligned with (orthogonal to) the identified direction. The energy gap of the detector is $\sigma\Omega = 0.75$, and $\mathbf{d}_A = \mathbf{d}_B$ and $|\mathbf{d}_A|/\sigma = 0.25$.

We observe that for detectors coupled to either twisted or untwisted fields with the chosen energy gap, the amount of entanglement that results in the final state depends on the orientation of the detectors with respect to the identified direction.

Fig. 4.9 The difference between the correlation function associated with the final state of two detectors in the cylindrical spacetime $\mathcal{M}_-$ and Minkowski space $\mathcal{M}$, $\mathrm{corr}(\rho_{AB})_{\mathcal{M}_-} - \mathrm{corr}(\rho_{AB})_{\mathcal{M}}$, is plotted as a function of their separation L/σ and energy gap $\sigma\Omega$, for detectors coupled to (**a**) untwisted fields ($\eta = 1$) and (**b**) twisted fields ($\eta = -1$). In both (**a**) and (**b**) the circumference of the spacetime is $a/\sigma = 4$, $\mathbf{d}_A = \mathbf{d}_B$, $|\mathbf{d}_A|/\sigma = 0.25$, and $\theta = 0$

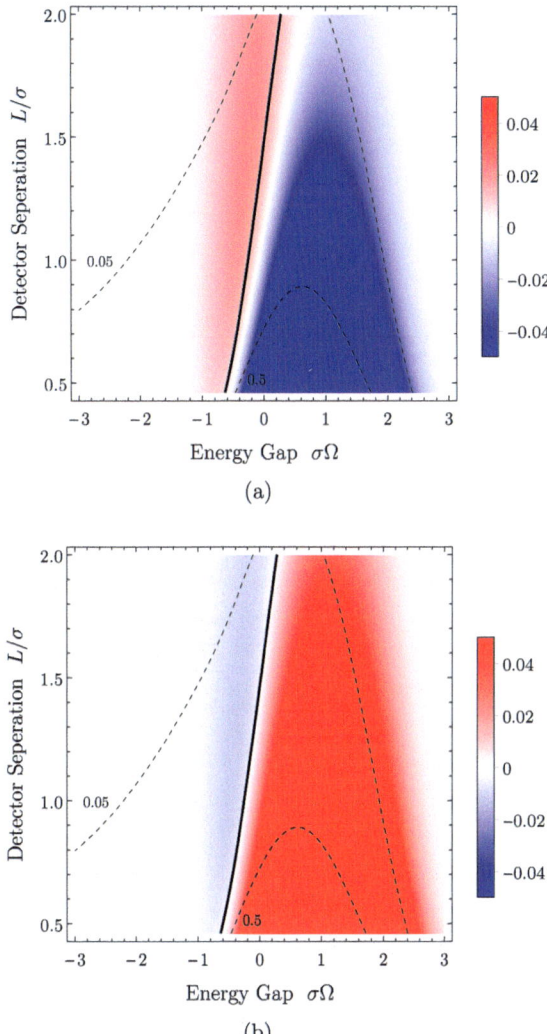

For large detector separation, the total amount of entanglement in the final state is less than for small detector separation; however, the dependence on the detector orientation is greater. For untwisted (twisted) fields and detectors with the chosen energy gap, the entanglement in the final state decreases (increases) as the angle with respect to the identified direction increases. This is opposite to the behaviour of detectors in $\mathcal{M}_0$, as illustrated in Fig. 4.7.

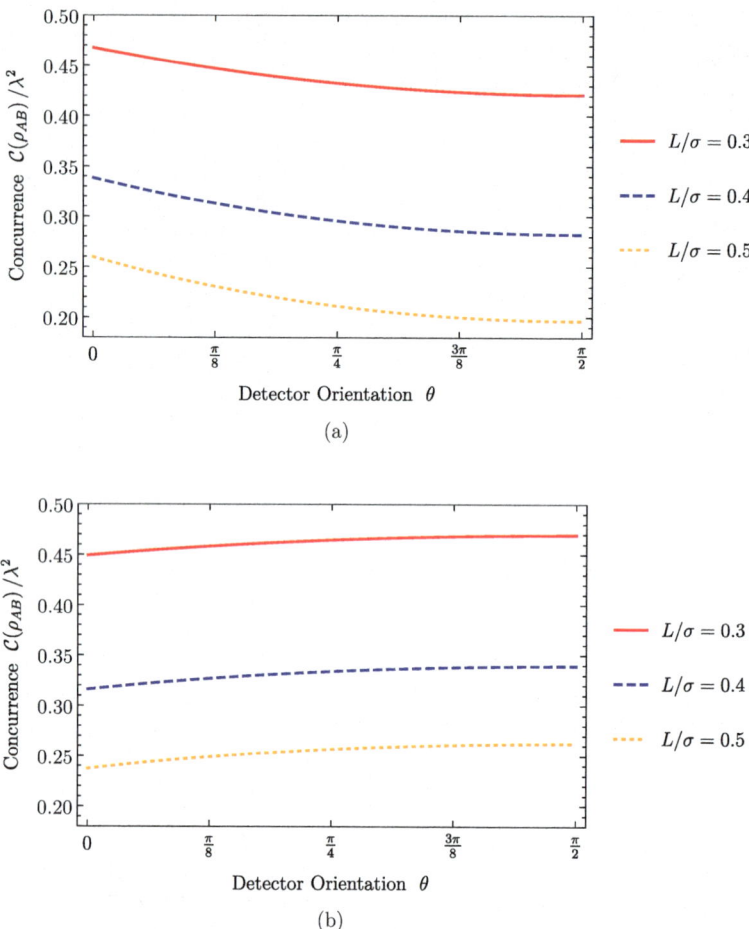

Fig. 4.10 For two detectors in the cylindrical spacetime $\mathcal{M}_-$, the concurrence associated with the final state of the two detectors $\mathcal{C}(\rho_{AB})_{\mathcal{M}_-}$ is plotted as a function of the detectors orientation with respect to the identified direction; $\theta = 0$ ($\theta = \pi/2$) corresponds to detectors aligned with (orthogonal to) the identified direction. The energy gap of the detectors is $\sigma\Omega = 0.75$. This is done for detectors coupled to (**a**) untwisted fields ($\eta = 1$) and (**b**) twisted fields ($\eta = -1$)

4.4 Summary

In summary, the behaviour of Unruh-DeWitt detectors depends on the global topology of the spacetime in which they live. This was demonstrated by examining detectors in three topologically distinct spacetimes, Minkowski space and the cylindrical spacetimes $\mathcal{M}_0$ and $\mathcal{M}_-$. This may be surprising given that these spacetimes are all locally indistinguishable and the detectors interact locally with the field.

However, the reason for the difference in the behaviour of detectors is that the vacuum state is different in all three spacetimes, and depends on the boundary conditions satisfied by the field—both the identifications used in constructing the cylindrical spacetimes $\mathcal{M}_0$ and $\mathcal{M}_-$ and whether the field is twisted or not. This is a quantum property of the vacuum state and would not occur classically because the classical vacuum state of the field vanishes everywhere and not subject to quantum fluctuations.

What has been shown in this chapter is that the transition probability of a detector interacting locally with the field and the entanglement and correlations harvested by a pair of detectors are sensitive to these boundary conditions. Specifically, both the correlations and entanglement harvested by a pair of detectors is greatest when their energy gap is small and positive. Furthermore, the orientation of a pair of detectors with respect to the identified direction in both $\mathcal{M}_0$ and $\mathcal{M}_-$ affects the entanglement harvested by the detectors.[4]

In the process of investigating how topological identifications of Minkowski space affect the entanglement harvesting protocol, the matrix elements P_A, P_B, X, and C defining the final state ρ_{AB} of a pair of detectors were computed in Minkowski space and the two cylindrical spacetimes $\mathcal{M}_0$, and $\mathcal{M}_-$; these quantities are summarized in Tables 4.1, 4.2, and 4.3.

The cylindrical spacetimes $\mathcal{M}_0$ and $\mathcal{M}_-$ studied in this chapter can equivalently be thought of as cavities with appropriate boundary conditions imposed on the field living inside the cavity. As discussed above, these boundary conditions affect how entangled a pair of detectors become while interacting with the vacuum state of the field. In a potential experiment to test for the presence of vacuum entanglement, one might construct a cavity in such a way that the resulting boundary conditions imposed on the field serve to amplify the entanglement harvested by a pair of detectors.

References

1. R. Banach, J.S. Dowker, Automorphic field theory-some mathematical issues. J. Phys. A **12**, 2527 (1979)
2. R. Banach, J.S. Dowker, The vacuum stress tensor for automorphic fields on some flat spacetimes. J. Phys. A **12**, 2545 (1979)
3. I.S. Gradshteyn, I. Ryzhik, *Table of Integrals, Series, and Products* (Academic, Cambridge, 1980)
4. P. Langlois, Imprints of Spacetime Topology in the Hawking-Unruh Effect. PhD thesis, University of Nottingham, 2005
5. A. Pozas-Kerstjens, E. Martín-Martínez, Harvesting correlations from the quantum vacuum. Phys. Rev. D **92**, 064042 (2015)

---

[4]One might imagine a Michelson-Morley-Bell experiment in which a large number of pairs of atoms in a cavity interact with the electromagnetic vacuum and become entangled. From these atoms, entanglement is distilled and used to violate a Bell inequality. Perhaps, one would find that the success of violating a Bell inequality would depend on the orientation of the atoms with respect to the cavity walls.

Chapter 5
Unruh-DeWitt Detectors Around (2+1)-Dimensional Black Holes

Quantum field theories have proven extremely useful in describing the fundamental interactions that govern our world—the Weinberg-Salam model has successfully unified the electromagnetic and weak interactions, and quantum chromodynamics provides an excellent description of the strong force. The success of these theories relies on perturbative quantum field theory. However, general relativity refuses to admit such a quantum description.

Constructing a consistent theory of quantum gravity is hard. It necessitates a radical departure from conventional quantum field theory—no longer can we quantize matter on a fixed background, everything must be quantized together including spacetime itself. Furthermore, taking general relativity as an ordinary field theory, the coupling constant has units of inverse mass, which implies the theory is nonrenormalizable and will fail to be perturbatively quantized. These issues, and others [12], pose difficult problems that need to be overcome in the construction of a satisfactory theory of quantum gravity.

Confronted with these issues, we are motivated to look for simpler models of gravity. General relativity in (2+1)-dimensions is such a model. Many of the fundamental issues with quantizing the (3+1)-dimensional theory appear in the (2+1)-dimensional theory. However, the (2+1)-dimensional theory is both mathematically and physically simpler. For example, the only degrees of freedom of the lower dimensional theory are topological and every solution to the field equations is either flat or has constant curvature. For these reasons, studying gravity in (2+1)-dimensions has been very instructive [5].

The Einstein-Hilbert action for a (2+1)-dimensional spacetime is

$$ S = \frac{1}{16\pi G} \int dx^3 \sqrt{-g} \, [R - 2\Lambda] , \tag{5.1} $$

where R is the Ricci scalar and Λ the cosmological constant. Variation of this action with respect to the metric yields the equations of motion for the gravitational degrees

© Springer Nature Switzerland AG 2019
A. R. H. Smith, *Detectors, Reference Frames, and Time*, Springer Theses,
https://doi.org/10.1007/978-3-030-11000-0_5

of freedom of the theory. For a negative cosmological constant $\Lambda = -1/\ell^2$ these equations admit a black hole solution of the form

$$ds^2 = -\left(\frac{r^2 - r_h^2}{\ell^2}\right) dt^2 + \left(\frac{r^2 - r_h^2}{\ell^2}\right)^{-1} dr^2 + r^2 d\phi^2, \qquad (5.2)$$

in Schwarzchild-like coordinates: $t \in (-\infty, \infty)$, $r \in (0, \infty)$, and $\phi \in (0, 2\pi)$. This solution has a horizon at $r_h = \ell\sqrt{M}$, where M is the mass of the black hole. This spacetime was discovered in 1992 by Bañados et al. [2, 3] and is appropriately known as the BTZ black hole; generalizations to charged and rotating black holes exist [5].

By making an appropriate topological identification of the BTZ black hole, the $\mathbb{RP}^2$ geon spacetime is constructed. The $\mathbb{RP}^2$ geon[1,2] is locally indistinguishable from the BTZ spacetime in the exterior region (the spacetime metric is the same). We will see that the $\mathbb{RP}^2$ geon black hole is an intermediate case between stationary and dynamical black holes, in the sense that the non-stationary features are behind the past and future horizons of the black hole [16].

The first aim of this chapter is to investigate how quantum field theory on the BTZ and $\mathbb{RP}^2$ geon black holes differ,[3] especially in their exterior regions where they are locally identical. To do so, we evaluate the transition rate of a detector sitting at a fixed distance away from the horizon of both black holes. We will demonstrate that a detector operating in the exterior region of the $\mathbb{RP}^2$ geon black hole develops a time-dependent transition rate, and is therefore sensitive to the non-stationary features behind its horizons [24].

The second aim of this chapter is to examine the entanglement harvesting protocol developed in Chap. 3 in the BTZ spacetime. In addition to serving as an example of the formalism, this will allow us to probe the entanglement structure of the Hartle-Hawking vacuum of a conformally coupled massless scalar field in an

---

[1]The term geon is short for "gravitational-electromagnetic entity", and was introduced by Wheeler [27] as a configuration of the gravitational field which has the spatial topology of $\mathbb{R}^3$ and is asymptotically flat, so that the mass of the geon may be defined by Arnowitt-Deser-Misner methods. Wheeler's goal was to describe all of classical physics in terms of these geons. To quote Misner and Wheeler [21]:

> If classical physics can be regarded as comprising gravitation, source free electromagnetism, unquantized charge, and unquantized mass of concentrations of electromagnetic field energy (geons), then classical physics can be described in terms of curved empty space, and nothing more.

Sorkin generalized this notion of the geon to non-trivial spatial topologies [25], allowing for the possibility of black hole geons, such as the $\mathbb{RP}^2$ geon we will consider here.

[2]As discussed by Louko [16], these black hole geons are not expected to be the result of stellar collapse as the non-trivial topology of the black hole geon is present since arbitrarily early times.

[3]Such an investigation was first carried out by Louko and Marolf [17] in the case of the Schwarzschild and associated $\mathbb{RP}^3$ geon, albeit with different methods and focus.

operational manner. We will investigate how the entanglement that results between two Unruh-DeWitt detectors interacting locally with the vacuum depends on the properties of BTZ black hole [9].

We begin this chapter in Sect. 5.1 by constructing the BTZ spacetime via topological identifications of (2+1)-dimensional anti-de Sitter space (AdS$_3$). Then we construct the $\mathbb{R}\mathbf{P}^2$ geon spacetime by further identification. In Sect. 5.2 we derive the Wightman functions associated with the Hartle-Hawking vacuum on both spacetimes from the AdS$_3$ Wightman functions using the method of images. In Sect. 5.3 we compare the transition rate of a stationary detector outside the BTZ horizon with an identical detector in the $\mathbb{R}\mathbf{P}^2$ geon spacetime. We find that while the transition rate is constant in the BTZ spacetime, the transition rate of the same detector in the $\mathbb{R}\mathbf{P}^2$ geon spacetime is time-dependent, even though the spacetime metric is identical in the region in which the detectors are operating. In Sect. 5.4 we examine the entanglement harvesting protocol for two detectors located outside the BTZ horizon and interpret the results in terms of the local Hawking temperature experienced by the detectors and red shift effects. We summarize the results present in Sect. 5.5.

5.1 The BTZ and $\mathbb{R}\mathbf{P}^2$ Geon Black Hole Spacetimes

In this section we will present the quotient space construction of the BTZ and $\mathbb{R}\mathbf{P}^2$ geon black holes. The (2+1)-dimensional AdS$_3$ space can be defined as the restriction to the submanifold

$$X_1^2 + X_2^2 - T_1^2 - T_2^2 = -\ell^2, \tag{5.3}$$

where $\ell > 0$ is the AdS length scale, embedded in the flat four-dimensional space $\mathbb{R}^{2,2}$ with coordinates (X_1, X_2, T_1, T_2) and metric

$$ds^2 = dX_1^2 + dX_2^2 - dT_1^2 - dT_2^2. \tag{5.4}$$

The BTZ black hole may be constructed by quotienting an open region of AdS$_3$ with the isometry group Z [5]. A set of coordinates on AdS$_3$ well suited to implement this quotient space construction and which cover the exterior region of the black hole are the BTZ coordinates:

$$X_1 = \ell \frac{r}{r_h} \sinh\left(\frac{r_h}{\ell}\phi\right), \qquad X_2 = \ell \sqrt{\frac{r^2}{r_h^2} - 1} \cosh\left(\frac{r_h}{\ell^2}t\right),$$

$$T_1 = \ell \frac{r}{r_h} \cosh\left(\frac{r_h}{\ell}\phi\right), \qquad T_2 = \ell \sqrt{\frac{r^2}{r_h^2} - 1} \sinh\left(\frac{r_h}{\ell^2}t\right), \tag{5.5}$$

where

$$t \in (-\infty, \infty), \qquad r \in (r_h, \infty), \qquad \text{and} \qquad \phi \in (-\infty, \infty). \tag{5.6}$$

In these coordinates the induced metric on AdS$_3$ from the embedding space $\mathbb{R}^{2,2}$ is

$$ds^2 = -\left(\frac{r^2 - r_h^2}{\ell^2}\right) dt^2 + \left(\frac{r^2 - r_h^2}{\ell^2}\right)^{-1} dr^2 + r^2 d\phi^2. \tag{5.7}$$

In the BTZ coordinates, the Z quotient is realized by the identification

$$\Gamma : (t, r, \phi) \sim (t, r, \phi + 2\pi), \tag{5.8}$$

so that $Z \simeq \{\Gamma^n\}$. The action of this quotient results in the coordinate ϕ becoming an angular coordinate $\phi \in (0, 2\pi)$. The resulting spacetime is known as the BTZ black hole $\mathcal{M}_{\text{BTZ}}$ and its metric is given by Eq. (5.7). Since the BTZ spacetime was constructed from a Z quotient of AdS$_3$, it is a quotient spacetime

$$\mathcal{M}_{\text{BTZ}} = \mathcal{M}_{\text{AdS}_3}/Z. \tag{5.9}$$

The $\mathbb{R}\mathbf{P}^2$ geon spacetime is constructed by a further quotient, which is best realized in the null coordinates U and V, which cover the entire BTZ spacetime:

$$\frac{r}{r_h} = \frac{1 - UV}{1 + UV} \qquad \text{and} \qquad \frac{r_h t}{\ell^2} = \ln \sqrt{-\frac{V}{U}}. \tag{5.10}$$

The embedding coordinates (X_1, X_2, T_1, T_2) may be expressed in terms of the null coordinates as

$$X_1 = \ell \left(\frac{1 - UV}{1 + UV}\right) \sinh\left(\frac{r_h}{\ell}\phi\right), \qquad X_2 = \ell \frac{V - U}{1 + UV},$$

$$T_1 = \ell \left(\frac{1 - UV}{1 + UV}\right) \cosh\left(\frac{r_h}{\ell}\phi\right), \qquad T_2 = \ell \frac{V + U}{1 + UV}. \tag{5.11}$$

In the null coordinates the BTZ metric takes the form

$$ds^2 = -\frac{4\ell^2}{(1 + UV)^2} dU dV + r_h^2 \left(\frac{1 - UV}{1 + UV}\right)^2 d\phi^2. \tag{5.12}$$

The $\mathbb{R}\mathbf{P}^2$ geon spacetime $\mathcal{M}_{\text{geon}}$ is constructed by the following identification of $\mathcal{M}_{\text{BTZ}}$ [18, 19]

$$J : (U, V, \phi) \sim (V, U, \phi + \pi). \tag{5.13}$$

As $\phi \in (0, 2\pi)$ is an angular coordinate we see that $J^2 = e$, and therefore J generates a $Z_2 \simeq \{e, J\}$ action on the BTZ spacetime. Therefore, the geon spacetime is seen to be the quotient spacetime

$$\mathcal{M}_{\text{geon}} = \mathcal{M}_{\text{BTZ}}/Z_2. \tag{5.14}$$

The Z_2 quotient acts without fixed points and properly discontinuously [16].

In terms of the Penrose diagram in Fig. 5.1a, the identification J acts by a reflection around the vertical axis and a rotation by π in the suppressed ϕ direction; this results in the Penrose diagram of the $\mathbb{R}\mathbf{P}^2$ geon spacetime depicted in Fig. 5.1b. In the BTZ coordinates J maps a point (t, r, ϕ) in region I to the point $(-t, r, \phi + \pi)$ in region III. In regions II and IV, J acts by identifying the points $(t, r, \phi) \sim (-t, r, \phi + \pi)$. As a result of these identifications, the spatial topology of the $\mathbb{R}\mathbf{P}^2$ geon spacetime is the real projective space[4] $\mathbb{R}\mathbf{P}^2/\{\text{point at infinity}\}$. These identifications also select a preferred spatial slice at $t = 0$ [16].

As mentioned in the introduction to this chapter, the $\mathbb{R}\mathbf{P}^2$ geon spacetime is an intermediate case between a stationary black hole and a dynamical black hole. A spacetime is stationary if there exists a globally defined timelike Killing field, otherwise the spacetime is dynamic [26]. One may expect that both the BTZ and

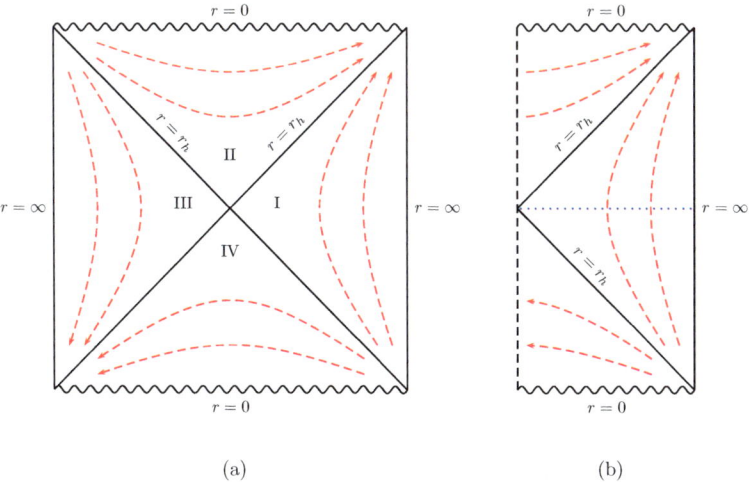

(a)   (b)

Fig. 5.1 The Penrose diagram of the (**a**) BTZ and (**b**) $\mathbb{R}\mathbf{P}^2$ geon black holes [3, 8]. The singularities are indicated by the wavy lines and the horizons by the diagonal lines. The red dashed lines indicated the orbits of the timelike Killing field ∂_t. In (**b**) the blue dotted line indicates the preferred time $t = 0$; see the discussion below Eq. (5.34)

---

[4]$\mathbb{R}\mathbf{P}^2$ is the topological space of lines passing through the origin in $\mathbb{R}^3$ and is best thought of as a 2-sphere with points on opposite sides of the sphere identified.

$\mathbb{RP}^2$ geon spacetimes admit such a Killing field seeing as in the exterior region both black holes are locally indistinguishable (their metric is the same), and expressed in the BTZ coordinates the metric is independent of the coordinate t. However, only the BTZ spacetime is stationary.

This is best illustrated by examining the orbits of the timelike Killing field ∂_t in the $\mathbb{RP}^2$ geon spacetime [16], which are plotted in red in Fig. 5.1b. Points along the dashed line behind the future horizon are identified with points along the dashed line behind the past horizon. As a consequence, where the orbits of the timelike Killing field run into the dashed line, the Killing field has a sign ambiguity. This results in the timelike Killing field ∂_t not being globally defined on the $\mathbb{RP}^2$ geon spacetime. This feature of the $\mathbb{RP}^2$ geon black hole is hidden behind its past and future horizons. However, as we will show in Sect. 5.3, an Unruh-DeWitt detector operating in the exterior region of the $\mathbb{RP}^2$ geon spacetime is sensitive to this feature.

5.2 The Wightman Function in the AdS$_3$, BTZ, and $\mathbb{RP}^2$ Geon Spacetimes

Beginning with the Wightman function in AdS$_3$, we derive the Wightman function in both the BTZ and $\mathbb{RP}^2$ geon spacetimes via the method of images. For a conformally coupled massless scalar field described by the action

$$I = -\int d^3x \sqrt{-g} \left(\frac{1}{2} g^{\mu\nu} \partial_\mu \phi \, \partial_\nu \phi + \frac{1}{16} R \phi^2 \right), \tag{5.15}$$

the Wightman function in AdS$_3$ is [1]

$$W_{\text{AdS}_3}(x, x') = \frac{1}{4\pi \sqrt{2}\ell} \left(\frac{1}{\sqrt{\sigma(x, x')}} - \frac{\zeta}{\sqrt{\sigma(x, x') + 2}} \right), \tag{5.16}$$

where $\sigma(x, x')$ is the square of the geodesic distance between the spacetime points x and x' in the embedding space $\mathbb{R}^{2,2}$ divided by ℓ^2

$$\sigma(x, x') = \frac{1}{2\ell^2} \left[(X_1 - X_1')^2 - (T_1 - T_1')^2 + (X_2 - X_2')^2 - (T_2 - T_2')^2 \right]. \tag{5.17}$$

The parameter $\zeta \in \{-1, 0, 1\}$ appearing in Eq. (5.16) specifies either Neumann ($\zeta = -1$), transparent ($\zeta = 0$), or Dirichlet ($\zeta = 1$) boundary conditions satisfied by the field at spatial infinity.

The BTZ Wightman function can be constructed from the AdS$_3$ Wightman function using the method of images[5]

$$W_{\mathrm{BTZ}}(x, x') = \sum_{n=-\infty}^{\infty} W_{\mathrm{AdS}_3}(x, \Gamma^n x')$$

$$= \frac{1}{4\pi\sqrt{2}\ell} \sum_{n=-\infty}^{\infty} \left(\frac{1}{\sqrt{\sigma(x, \Gamma^n x')}} - \frac{\zeta}{\sqrt{\sigma(x, \Gamma^n x') + 2}} \right). \qquad (5.18)$$

The evaluation of $\sigma(x, \Gamma^n x')$ depends on whether the points x and x' are inside or outside the horizon of the black hole. We will ultimately be interested in detectors outside the black hole, so let us suppose $x = (t, r, \phi)$ and $x' = (t', r', \phi')$ with $r, r' > r_h$. To evaluate $\sigma(x, \Gamma^n x')$, Eq. (5.5) is substituted into Eq. (5.17) yielding

$$\sigma(x, \Gamma^n x') = \frac{rr'}{r_h^2} \cosh\left[\frac{r_h}{\ell}(\Delta\phi - 2\pi n)\right] - 1$$

$$- \frac{(r^2 - r_h^2)^{\frac{1}{2}}(r'^2 - r_h^2)^{\frac{1}{2}}}{r_h^2} \cosh\left[\frac{r_h}{\ell^2}\Delta t\right], \qquad (5.19)$$

where $\Delta\phi := \phi - \phi'$ and $\Delta t := t - t'$.

From Eqs. (5.18) and (5.19) we note that $W_{\mathrm{BTZ}}(x, x')$ is periodic in imaginary time with a period

$$T^{-1} = \frac{2\pi\ell^2}{r_h} = \frac{2\pi}{\kappa}, \qquad (5.20)$$

where κ is the surface gravity of the black hole. This periodicity implies that $W_{\mathrm{BTZ}}(x, x')$ is a thermal Wightman function associated with the temperature T [5]. Further, by examining the analyticity properties of $W_{\mathrm{BTZ}}(x, x')$, Lifschytz and Ortiz [14] demonstrated that this Wightman function is associated with the Hartle-Hawking vacuum.[6]

The $\mathbb{R}\mathbf{P}^2$ geon Wightman function can be constructed by an additional image sum over $Z_2 \simeq \{e, J\}$

$$W_{\mathrm{geon}}(x, x') = W_{\mathrm{BTZ}}(x, x') + W_{\mathrm{BTZ}}(x, Jx'), \qquad (5.21)$$

where

---

[5]We will restrict ourselves to the study of untwisted fields and set $\eta = 1$ in Eq. (B.1).
[6]The Hartle-Hawking vacuum is the quantum state of the field that describes the equilibrium of the black hole with thermal radiation at the Hawking temperature $T = \kappa/2\pi$ [4, 22].

$$W_{\text{BTZ}}(x, Jx') = \sum_{n=-\infty}^{\infty} W_{\text{AdS}_3}(x, J\Gamma^n x')$$

$$= \frac{1}{4\pi\sqrt{2}\ell} \sum_{n=-\infty}^{\infty} \left(\frac{1}{\sqrt{\sigma(x, J\Gamma^n x')}} - \frac{\zeta}{\sqrt{\sigma(x, J\Gamma^n x') + 2}} \right).$$

$$(5.22)$$

This is the $\mathbb{R}P^2$ geon Wightman function induced by the BTZ Hartle-Hawking vacuum via the method of images. This Wightman function has been shown to correspond to a Hartle-Hawking-like vacuum state in the $\mathbb{R}P^2$ geon spacetime [8, 17].

In evaluating $\sigma(x_D(\tau), J\Gamma^n x_D(\tau - \ell\tilde{s}))$, we note that the action of J is to swap U and V and identify ϕ with $\phi + \pi$. Swapping U and V results in $t \to -t$ and $X_2 \to -X_2$, which can be seen from Eqs. (5.10) and (5.11), respectively. Again supposing $x = (t, r, \phi)$ and $x' = (t', r', \phi')$ and $r, r' > r_h$, $\sigma(x, J\Gamma^n x')$ evaluates to

$$\sigma(x, J\Gamma^n x') = \frac{rr'}{r_h^2} \cosh\left[\frac{r_h}{\ell}\left(\Delta\phi - 2\pi\left[n + \tfrac{1}{2}\right]\right)\right] - 1$$

$$+ \frac{(r^2 - r_h^2)^{\frac{1}{2}} (r'^2 - r_h^2)^{\frac{1}{2}}}{r_h^2} \cosh\left[\frac{r_h}{\ell^2}(t + t')\right]. \qquad (5.23)$$

From Eq. (5.23), together with Eqs. (5.21) and (5.22), we observe that the $\mathbb{R}P^2$ geon Wightman function no longer depends on the difference in coordinate times t and t', but rather on their sum. This implies that the Wightman function induced by the BTZ Wightman function via the method of images is not invariant under the isometry generated by the timelike Killing field ∂_t. This might have been expected based on the discussion at the end of Sect. 5.1, in which it was shown that the $\mathbf{R}P^2$ geon spacetime is not stationary. This feature is reflected in the time dependence of the $\mathbb{R}P^2$ geon Wightman function. For spacetime points in the far future or far past ($|t + t'| \to \infty$), we see from Eq. (5.23) that $W_{\text{BTZ}}(x, Jx')$ vanishes and the $\mathbf{R}P^2$ geon Wightman function is identical to the BTZ Wightman function. Further discussion of quantum field theory on geon spacetimes can be found in [15–19].

5.3   Detectors Outside the BTZ and $\mathbb{R}P^2$ Geon Black Holes

Using the Wightman functions derived in the previous section we will examine the behaviour of static Unruh-DeWitt detectors operating in the exterior region of both the BTZ and $\mathbf{R}P^2$ geon black holes.[7] Specifically, we will evaluate the transition

---

[7]Detectors in the BTZ spacetime have been studied in the past by Lifschytz and Ortiz [14] and more recently by Hodgkinson and Louko [10, 11].

rate of these detectors given in Eq. (3.16). We will find that while the transition rate of a detector outside the BTZ black hole is time-independent—as expected since the BTZ spacetime is stationary and the detector static—the transition rate of an equivalent detector in the $\mathbb{R}P^2$ geon spacetime is time-dependent [24].

In both the BTZ and $\mathbb{R}P^2$ geon spacetimes, suppose the trajectory of the detector is such that it remains at a fixed distance $R > r_h$ from the black hole

$$x_D(\tau) := \left\{ t = \frac{\ell}{\sqrt{R^2 - r_h^2}} \, \tau, \quad r = R, \quad \phi = \Phi, \right\}, \tag{5.24}$$

where the above trajectory is parametrized in terms of the detectors proper time τ.

We begin by evaluating the transition rate of a detector in the BTZ spacetime moving along the trajectory given in Eq. (5.24) with an energy gap Ω. In the sharp switching limit (discussed in Sect. 3.1), the transition rate is given by Eq. (3.16)

$$\dot{P}_{\text{BTZ}}(\tau) = \lambda^2 \left(\frac{1}{4} + 2 \int_0^{\Delta\tau} ds \; \text{Re} \left[e^{-i\Omega s \ell} W_{\text{BTZ}}(x_D(\tau), x_D(\tau - \ell\tilde{s})) \right] \right)$$

$$= \lambda^2 \left(\frac{1}{4} + \frac{1}{2\pi\sqrt{2}} \sum_{n=-\infty}^{\infty} \int_0^{\Delta\tau/\ell} d\tilde{s} \; \text{Re} \right.$$

$$\times \left[e^{-i\Omega\tilde{s}\ell} \left(\frac{1}{\sqrt{\sigma(x_D(\tau), \Gamma^n x_D(\tau - \ell\tilde{s}))}} \right. \right.$$

$$\left. \left. \left. - \frac{\zeta}{\sqrt{\sigma(x_D(\tau), \Gamma^n x_D(\tau - \ell\tilde{s})) + 2}} \right) \right] \right), \tag{5.25}$$

where we have introduced the dimensionless integration variable $\tilde{s} := s/\ell$. Further, we have

$$\sigma\left(x_D(\tau), \Gamma^n x_D(\tau - \ell\tilde{s})\right) = 2\frac{R^2 - r_h^2}{r_h^2} \left[\frac{R^2}{R^2 - r_h^2} \sinh^2\left(n\pi \frac{r_h}{\ell}\right) \right.$$

$$\left. - \sinh^2\left(\frac{r_h}{\sqrt{R^2 - r_h^2}} \frac{\tilde{s}}{2}\right) \right]. \tag{5.26}$$

As shown by Hodgkinson and Louko [10], and summarized in Appendix C, taking the limit in which the detector is turned on in the far past ($\Delta\tau \to \infty$) the transition rate can be written as

$$\dot{P}_{\text{BTZ}} = \frac{\lambda^2}{2\pi} e^{-\ell\beta\Omega/2} \sum_{n=-\infty}^{\infty} \int_0^{\infty} dy \cos\left(y\ell\beta\Omega/\pi\right)$$

$$\times \left[\frac{1}{\sqrt{K_n + \cosh^2 y}} - \frac{\zeta}{\sqrt{Q_n + \cosh^2 y}} \right], \tag{5.27}$$

where

$$K_n := \frac{R^2}{R^2 - r_h^2} \sinh^2\left(n\pi \frac{r_h}{\ell}\right), \quad Q_n := K_n + \frac{r_h^2}{R^2 - r_h^2}, \text{ and } \beta := 2\pi \frac{\sqrt{R^2 - r_h^2}}{r_h}.$$

Expressing the transition rate in this form lends itself to being evaluated numerically. Note the transition rate is independent of τ, and therefore we have dropped the explicit time dependence in Eq. (5.27), i.e. $\dot{P}_{\text{BTZ}}(\tau) \rightarrow \dot{P}_{\text{BTZ}}$.

As shown in [6, 7], the KMS condition[8] implies the following condition on the transition rate $\dot{P}_D(\Omega)$ of a particle detector with energy gap Ω

$$\dot{P}_D(\Omega) = e^{-\alpha\Omega} \dot{P}_D(-\Omega). \tag{5.29}$$

We note from Eq. (5.27) that the BTZ transition rate $\dot{P}_{\text{BTZ}}(\Omega)$ satisfies the KMS condition with $\alpha = \ell\beta$. Therefore the detector sees the field at a temperature $1/\ell\beta$, which corresponds to the local Hawking temperature[9] at the location of the detector [10]. This conclusion will be important for the interpretation of the results in the next section.

We now evaluate the transition rate of the same detector outside the $\mathbb{RP}^2$ geon black hole. Since the $\mathbb{RP}^2$ geon Wightman function is equal to the Wightman function in the BTZ spacetime plus an image term, Eq. (5.21), the transition rate in the $\mathbb{RP}^2$ geon spacetime is

$$\dot{P}_{\text{geon}}(\tau) = \dot{P}_{\text{BTZ}} + \Delta\dot{P}(\tau), \tag{5.30}$$

where

---

[8]The Kubo-Martin-Schwinger (KMS) condition [13, 20] is a general definition of equilibrium states in terms of the Wightman function. For a timelike trajectory $x(\tau)$, a state of the field $\rho \in \mathcal{S}(\mathcal{H}_\phi)$ satisfies the KMS condition if the Wightman function $W_\rho(\tau, \tau') := \text{tr}\left(\rho \, \phi[x(\tau)] \, \phi[x(\tau')]\right)$ satisfies

$$W(\tau - i\alpha, \tau') = W(\tau', \tau), \tag{5.28}$$

the temperature of the state being $1/\alpha$.
[9]The local Hawking temperature is given by the Tolman relation $1/\ell\beta = (-g_{00})^{-1/2} T$, where g_{00} is the coefficient in front of dt^2 in the line element in Eq. (5.2).

$$\Delta\dot{P}(\tau) := \frac{1}{2\pi\sqrt{2}} \sum_{n=-\infty}^{\infty} \int_0^{\infty} d\tilde{s} \, \mathrm{Re}\left[e^{-i\Omega\tilde{s}\ell}\left(\frac{1}{\sqrt{\sigma(x_D(\tau), J\Gamma^n x_D(\tau - \ell\tilde{s}))}}\right.\right.$$

$$\left.\left. - \frac{\zeta}{\sqrt{\sigma(x_D(\tau), J\Gamma^n x_D(\tau - \ell\tilde{s})) + 2}}\right)\right], \tag{5.31}$$

and

$$\sigma(x_D(\tau), J\Gamma^n x_D(\tau - \ell\tilde{s})) = \frac{\beta^2}{2\pi^2}\left(\frac{R^2}{R^2 - r_h^2}\sinh^2\left[\pi\left(n + \tfrac{1}{2}\right)\frac{r_h}{\ell}\right]\right.$$

$$\left. + \cosh^2\left[\frac{\pi\tau}{\beta\ell} - \frac{\pi\tilde{s}}{\beta}\right]\right). \tag{5.32}$$

Defining

$$\bar{K}_n := \frac{R^2}{R^2 - r_h^2}\sinh^2\left[\pi\left(n + \tfrac{1}{2}\right)\frac{r_h}{\ell}\right] \quad \text{and} \quad \bar{Q}_n := K_n + \frac{r_h^2}{R^2 - r_h^2}, \tag{5.33}$$

and changing the integration variable to $y = \pi\tilde{s}/\beta$, the second term in Eq. (5.30) becomes

$$\Delta\dot{P}(\tau) := \frac{1}{2\pi} \sum_{n=-\infty}^{\infty} \int_0^{\infty} dy \, \cos(\Omega\ell\beta y/\pi) \left[\frac{1}{\sqrt{\bar{K}_n + \cosh^2\left[y - \frac{\pi\tau}{\beta\ell}\right]}}\right.$$

$$\left. - \frac{\zeta}{\sqrt{\bar{Q}_n + \cosh^2\left[y - \frac{\pi\tau}{\beta\ell}\right]}}\right]. \tag{5.34}$$

From Eq. (5.34) we see that for large negative τ the integrand is small and $\Delta\dot{P}(\tau) \approx 0$. With this observation and Eq. (5.30), we conclude that the transition rate of a detector operating in the far past (large negative τ) is identical to the transition rate of the same detector in the BTZ spacetime. For $\tau \gtrsim 0$, the integrand does not vanish and $\Delta\dot{P}(\tau)$ is significant. From these observations we conclude that the time-dependent contribution to the $\mathbb{R}\mathbf{P}^2$ geon transition rate $\Delta\dot{P}(\tau)$ turns on around $\tau \approx 0$, and remains on as $\tau \to \infty$. At this proper time ($\tau = 0$), the detector is on the preferred hypersurface $t = 0$, which is singled out by the non-stationary features of the $\mathbb{R}\mathbf{P}^2$ geon spacetime located behind its past and future horizons (see Fig. 5.1). The fact that the transition rate develops a time-dependence after the detector has crossed this surface ($t = 0$) demonstrates the detectors dependence on the non-stationary features of the $\mathbb{R}\mathbf{P}^2$ geon spacetime.

Having simplified the expressions for the transition rate of a detector in the BTZ and $\mathbb{RP}^2$ geon spacetimes in Eqs. (5.27), (5.30), and (5.34), we now evaluate these expressions numerically[10] and plot these transition rates as a function of the read out time of the detector and the detector's energy gap for the field satisfying Neumann (Fig. 5.2), transparent (Fig. 5.3), and Dirichlet boundary conditions (Fig. 5.4).

A few remarks on the plots shown in Figs. 5.2, 5.3, and 5.4 are given here.

1. The first observation to be made from Figs. 5.2, 5.3, and 5.4 or Eq. (5.34) is that the transition rate of a detector in the $\mathbf{R}\mathbf{P}^2$ geon spacetime is time-dependent. That is, a detector operating in the exterior region of the $\mathbb{RP}^2$ geon spacetime is sensitive to the non-stationary features of the spacetime located behind its horizon. This time dependence appears after the detector has crossed the preferred hypersurface $t = 0$, after which the transition rate oscillates around the transition rate of an identical detector in the BTZ spacetime. This suggests that, in principle, by observing the transition rate of such a detector one may be able to infer the location of this preferred time slice.

2. In the analysis above the detector was turned on in the asymptotic past. From Figs. 5.2a, 5.3a, and 5.4a we see that the transition rate of a detector in the $\mathbb{RP}^2$ geon spacetime only differs from the transition rate of a detector in the BTZ spacetime when the read out time is approximately greater or equal to the preferred time $t = 0$. This behaviour was observed in [17, 18] for detectors in the $\mathbb{RP}^3$ geon spacetime.

3. From Figs. 5.2b, 5.3b, and 5.4b we see that detectors in the $\mathbf{R}\mathbb{P}^2$ geon spacetime with an energy gap close to zero vary more dramatically than detectors with a larger or smaller energy gap. This could have been anticipated by noting that the integrand in Eq. (5.34) oscillates with a frequency proportionally to Ω. We see that by tuning Ω to a value close to zero increases the detector's sensitivity to the non-stationary features of the $\mathbf{R}\mathbb{P}^2$ geon spacetime.

4. Upon comparison of Figs. 5.2, 5.3, and 5.4 we see that the boundary condition satisfied by the field at spatial infinity affects the transition rate of a detector. The transition rate in both spacetimes is largest when the field satisfies Neumann ($\zeta = -1$) boundary conditions and smallest when it satisfies Dirichlet ($\zeta = 1$) boundary conditions.

---

[10]The numerical calculations were carried out in Mathematica. The integrals appearing in Eqs. (5.27) and (5.34) were evaluated from zero to infinity using `NIntegrate` with `PrecisionGoal` $\to$ 4, `AccuracyGoal` $\to$ 4, and `Method` $\to$ `"ExtrapolatingOscillatory"`. The sums appearing in Eqs. (5.27) and (5.34) were evaluated from $n = -20$ to $n = 20$. To generate the plots in Figs. 5.2, 5.3, and 5.4, the transition rate was evaluated for 80 and 120 points, respectively, uniformly sampled across the domain of each plot.

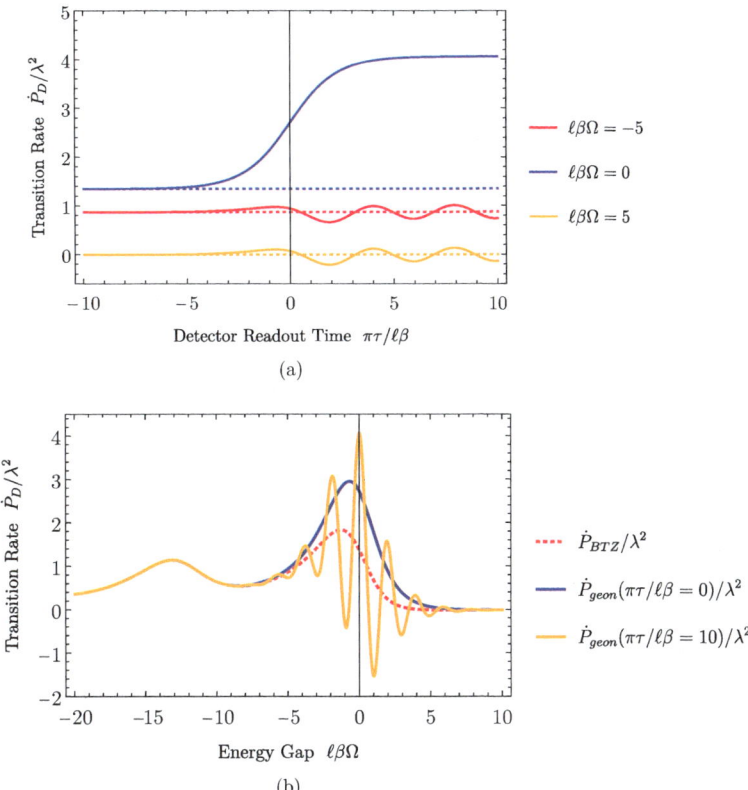

Fig. 5.2 The transition rate of a static detector in both the BTZ and $\mathbb{RP}^2$ geon spacetimes is plotted as a function of (**a**) the proper time $\pi\tau/\ell\beta$ at which the detector is read for a fixed energy gap of the detector, and (**b**) the energy gap of the detector $\ell\beta\Omega$ for a fixed proper time at which the detector is read. In both (**a**) and (**b**) $r_h/\ell = 0.5$, $R^2/r_h^2 = 10$, and the field satisfies Neumann boundary conditions ($\zeta = -1$) at spatial infinity. In (**a**) the dotted and solid lines of the same colour indicate the transition rate of an identical detector in the BTZ and $\mathbb{RP}^2$ geon spacetimes, respectively

5.4   Entanglement from the Black Hole Vacuum

How entangled do two static Unruh-DeWitt detectors become by interacting locally with the Hartle-Hawking vacuum in the exterior region of the BTZ black hole? To answer this question we will apply the entanglement harvesting protocol developed in Chap. 3 [9].

We consider two detectors A and B at fixed distances R_A and R_B from the BTZ black hole with identical energy gaps $\Omega = \Omega_A = \Omega_B$. The spacetime trajectories of such detectors are

$$x_A(\tau_A) := \{t = \tau_A/b_A, \quad r = R_A, \quad \phi = \Phi_A\}, \tag{5.35a}$$

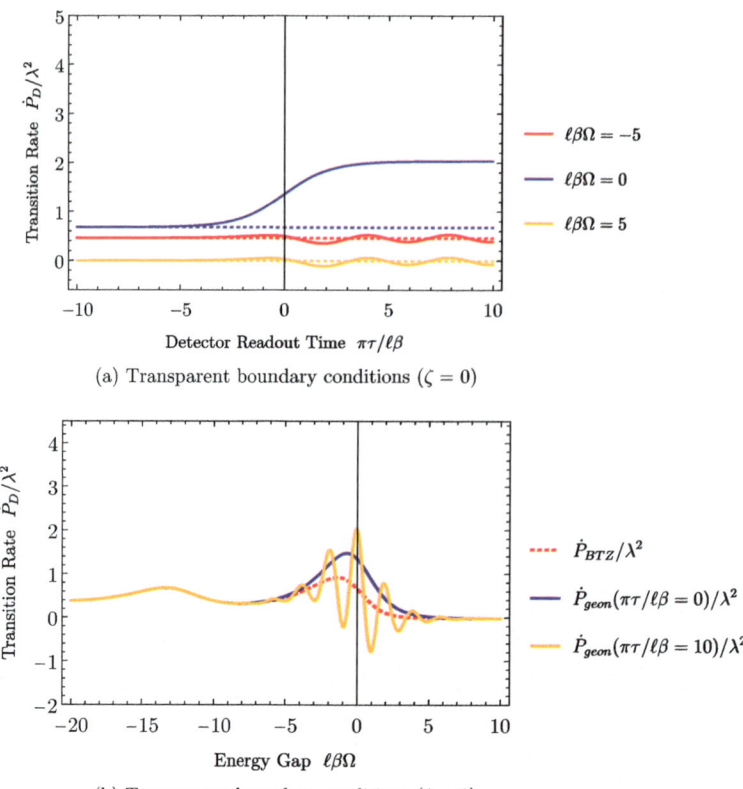

(a) Transparent boundary conditions ($\zeta = 0$)

(b) Transparent boundary conditions ($\zeta = 0$)

Fig. 5.3 The transition rate of a static detector in both the BTZ and $\mathbb{R}\mathbf{P}^2$ geon spacetimes is plotted as a function of (**a**) the proper time $\pi \tau/\ell\beta$ at which the detector is read for a fixed energy gap of the detector, and (**b**) the energy gap of the detector $\ell\beta\Omega$ for a fixed proper time at which the detector is read. In both (**a**) and (**b**) $r_h/\ell = 0.5$, $R^2/r_h^2 = 10$, and the field satisfies Neumann boundary conditions ($\zeta = 0$) at spatial infinity. In (**a**) the dotted and solid lines of the same colour indicate the transition rate of an identical detector in the BTZ and $\mathbb{R}\mathbf{P}^2$ geon spacetimes, respectively

$$x_B(\tau_B) := \{t = \tau_B/b_B, \quad r = R_B, \quad \phi = \Phi_B\}, \tag{5.35b}$$

where τ_A and τ_B are the proper time of each detector and b_A and b_B are red shift factors given by

$$b_A := \frac{d\tau_A}{dt} = \frac{\sqrt{R_A^2 - r_h^2}}{\ell} \quad \text{and} \quad b_B := \frac{d\tau_B}{dt} = \frac{\sqrt{R_B^2 - r_h^2}}{\ell}.$$

Without loss of generality, we will consider detector A being closer to the horizon than detector B, $r_h < R_A < R_B$.

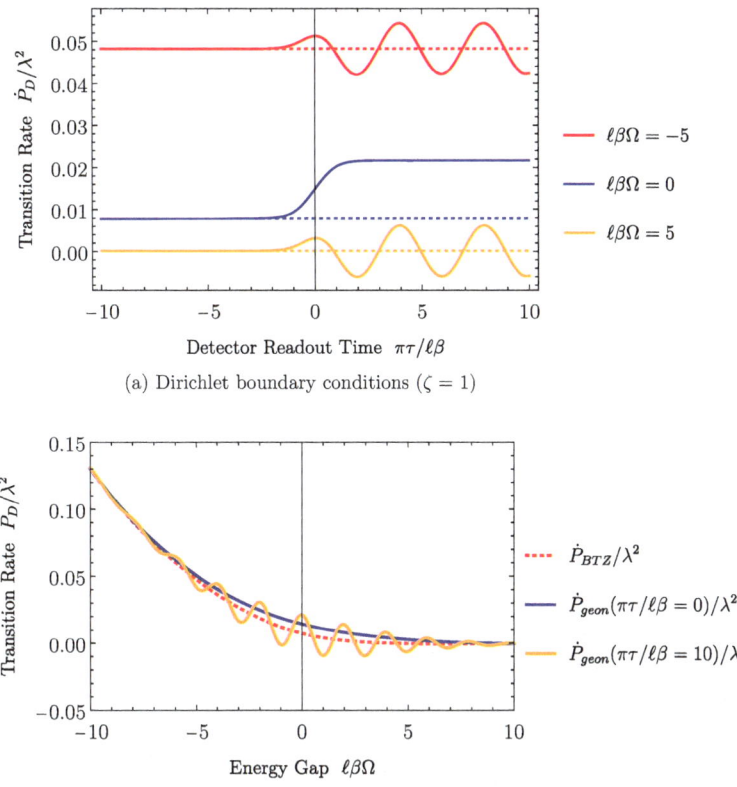

(a) Dirichlet boundary conditions ($\zeta = 1$)

(b) Dirichlet boundary conditions ($\zeta = 1$)

Fig. 5.4 The transition rate of a static detector in both the BTZ and $\mathbb{R}\mathbf{P}^2$ geon spacetimes is plotted as a function of (**a**) the proper time $\pi\tau/\ell\beta$ at which the detector is read for a fixed energy gap of the detector, and (**b**) the energy gap of the detector $\ell\beta\Omega$ for a fixed proper time at which the detector is read. In both (**a**) and (**b**), $r_h/\ell = 0.5$, $R^2/r_h^2 = 10$, and the field satisfies Neumann boundary conditions ($\zeta = 1$) at spatial infinity. In (**a**) the dotted and solid lines of the same colour indicate the transition rate of an identical detector in the BTZ and $\mathbb{R}\mathbf{P}^2$ geon spacetimes, respectively

As we did in Chap. 4, we will choose the switching functions of the detectors to be Gaussian

$$\chi_A(\tau_A) = e^{-\tau_A^2/2\sigma^2} \qquad \text{and} \qquad \chi_B(t) = e^{-t_B^2/2\sigma^2}, \tag{5.36}$$

with the interpretation that detector A and B are interacting with the field for an approximate amount of proper time $k\sigma$, centred around the spacelike hypersurface $t = 0$; k should be chosen so that the interaction between the field and detectors is negligible at the proper time $\pm k\sigma$.

Suppose that prior to the interaction with the field both detectors are in their ground state. After the interaction has ceased, the bipartite state of the detectors is given by Eq. (3.24)

$$
\rho_{AB} = \begin{pmatrix} 1 - P_A - P_B & 0 & 0 & X \\ 0 & P_B & C & 0 \\ 0 & C^* & P_A & 0 \\ X^* & 0 & 0 & 0 \end{pmatrix} + \mathcal{O}\left(\lambda^4\right).
\tag{5.37}
$$

To quantify how entangled the detectors become as a result of interacting with the field, we will make use of the concurrence $\mathcal{C}(\rho_{AB})$ given in Eq. (3.42)

$$
\mathcal{C}(\rho_{AB}) = 2 \max\left[0, \ |X| - \sqrt{P_A P_B} \right] + \mathcal{O}\left(\lambda^4\right),
\tag{5.38}
$$

where

$$
P_A = \lambda^2 \int d\tau d\tau' \, e^{-\tau^2/2\sigma^2} e^{-\tau'^2/2\sigma^2} e^{-i\Omega(\tau-\tau')} W_{\mathrm{BTZ}}\big(x_A(\tau), x_A(\tau')\big),
\tag{5.39a}
$$

$$
P_B = \lambda^2 \int d\tau d\tau' \, e^{-\tau^2/2\sigma^2} e^{-\tau'^2/2\sigma^2} e^{-i\Omega(\tau-\tau')} W_{\mathrm{BTZ}}\big(x_B(\tau), x_B(\tau')\big),
\tag{5.39b}
$$

$$
X = -\lambda^2 b_A b_B \int_{t>t'} dt dt' \Big[e^{-b_B^2 t^2/2\sigma^2} e^{-b_A^2 t'^2/2\sigma^2} e^{-i\Omega[b_B t + b_A t']}
$$
$$
\times W_{\mathrm{BTZ}}\big(x_A(t'), x_B(t)\big)
$$
$$
+ e^{-b_A^2 t^2/2\sigma^2} e^{-b_B^2 t'^2/2\sigma^2} e^{-i\Omega[b_A t + b_B t']} W_{\mathrm{BTZ}}\big(x_B(t'), x_A(t)\big) \Big],
\tag{5.39c}
$$

as given by Eqs. (3.9) and (3.26).

We now evaluate P_D for $D \in \{A, B\}$ by changing integration variables to $y := \tau - \tau'$ and $y' := \tau + \tau'$, and carrying out the integration over y', which results in

$$
P_D = \lambda^2 \sqrt{\pi} \sigma \int dy \, e^{-y^2/4\sigma^2} g(y) e^{-i\Omega y},
\tag{5.40}
$$

where $g(y) := W_{\mathrm{BTZ}}\big(x_A(\tau), x_A(\tau')\big)$, which from Eqs. (5.18) and (5.19) are seen to be a function of $y = \tau - \tau'$. From Eq. (5.40), note that P_D is the Fourier transform of the product of $g(y)$ and $e^{-y^2/4\sigma^2}$. Therefore, using the convolution theorem, we may express P_D as

$$
P_D = \lambda^2 \sigma^2 \int d\omega \, e^{-\sigma^2(\omega-\Omega)^2} F(\omega),
\tag{5.41}
$$

where

$$F(\omega) := \int dy\, e^{-i\omega y} g(y)$$

$$= \frac{1}{2} \frac{1}{e^{\omega \ell \beta} + 1} \sum_n \left[P_{\frac{i\omega \ell \beta}{2\pi} - \frac{1}{2}} \left(\cosh \alpha_n^- \right) - \zeta\, P_{\frac{i\omega \ell \beta}{2\pi} - \frac{1}{2}} \left(\cosh \alpha_n^+ \right) \right],$$

$$(5.42)$$

and P_ν is the Legendre function of the first kind and

$$\cosh \alpha_n^{\mp} := \frac{1}{b_D^2} \frac{r_h^2}{\ell^2} \left[\frac{R_D^2}{r_h^2} \cosh\left(2\pi n \frac{r_h}{\ell} \right) \mp 1 \right]. \qquad (5.43)$$

The integral defining $F(\omega)$ was first evaluated by Lifschytz and Ortiz [14]. Having expressed the transition probability P_D as a convolution of the Fourier transform of the switching function and $F(\omega)$, P_D may now be evaluated numerically.

We turn our attention to the evaluation of X in Eq. (5.39c). Using Eqs. (5.18) and (5.19) it is seen that X may be expressed as

$$X = -\lambda^2 \frac{b_A b_B}{4\pi \sqrt{2\ell}} \sum_{n=-\infty}^{\infty} \left(\left[I_n^-(A, B) + I_n^-(B, A) \right] - \zeta \left[I_n^+(A, B) + I_n^+(B, A) \right] \right),$$

$$(5.44)$$

where

$$I_n^{\mp}(A, B) := \frac{1}{\sqrt{b_A b_B}} \frac{r_h}{\ell} \int_{-\infty}^{\infty} dt \int_{-\infty}^{t} dt'\, \frac{e^{-b_B^2 t^2 / 2\sigma^2} e^{-b_A^2 t'^2 / 2\sigma^2} e^{-i\Omega[b_B t + b_A t']}}{\sqrt{\cosh \Xi_n^{\mp} - \cosh\left[\frac{r_h}{\ell^2}(t' - t) \right]}},$$

$$(5.45)$$

$$\cosh \Xi_n^{\mp} := \frac{1}{b_A b_B} \frac{r_h^2}{\ell^2} \left[\frac{R_A R_B}{r_h^2} \cosh\left(2\pi n \frac{r_h}{\ell} \right) \mp 1 \right]. \qquad (5.46)$$

To simplify the integrals in $I_n^{\mp}(A, B)$, let us change integration variables to $u := t$ and $s := t - t'$, so that

$$I_n^{\mp}(A, B) = \frac{1}{\sqrt{b_A b_B}} \frac{r_h}{\ell} \int_{-\infty}^{\infty} du$$

$$\times \int_{0}^{\infty} ds\, \frac{e^{-b_B^2 u^2 / 2\sigma^2} e^{-b_A^2 (u-s)^2 / 2\sigma^2} e^{-i\Omega[b_B u + b_A(u-s)]}}{\sqrt{\cosh \Xi_n^{\mp} - \cosh\left[\frac{r_h}{\ell^2} s \right]}}$$

$$= \frac{1}{\sqrt{b_A b_B}} \frac{r_h}{\ell} \int_0^\infty ds \, \frac{e^{-b_A^2 s^2/2\sigma^2} e^{i\Omega b_A s}}{\sqrt{\cosh \Xi_n^\mp - \cosh\left[\frac{r_h}{\ell^2} s\right]}}$$

$$\times \int_{-\infty}^\infty du \, e^{b_A^2 us/\sigma^2} e^{-(b_A^2 + b_B^2)u^2/2\sigma^2} e^{-i\Omega(b_A + b_B)u}$$

$$= \frac{\sqrt{2\pi}\sigma}{\sqrt{b_A b_B}\sqrt{b_A^2 + b_B^2}} \frac{r_h}{\ell} e^{-\frac{(b_A + b_B)^2}{b_A^2 + b_B^2}\frac{\Omega^2 \sigma^2}{2}}$$

$$\times \int_0^\infty ds \, \frac{e^{-\frac{b_A^2 b_B^2}{b_A^2 + b_B^2}\frac{s^2}{2\sigma^2}} e^{i\frac{b_A b_B}{b_A^2 + b_B^2}(b_B - b_A)\Omega s}}{\sqrt{\cosh \Xi_n^\mp - \cosh\left[\frac{r_h}{\ell^2} s\right]}}. \tag{5.47}$$

Observe that

$$I_n^\mp(A, B) + I_n^\mp(B, A) = \frac{2\sqrt{2\pi}\sigma}{\sqrt{b_A b_B}\sqrt{b_A^2 + b_B^2}} \frac{r_h}{\ell} e^{-\frac{(b_A + b_B)^2}{b_A^2 + b_B^2}\frac{\Omega^2 \sigma^2}{2}}$$

$$\times \int_0^\infty ds \, \frac{e^{-\frac{b_A^2 b_B^2}{b_A^2 + b_B^2}\frac{s^2}{2\sigma^2}} \cos\left[\frac{b_A b_B}{b_A^2 + b_B^2}(b_B - b_A)\Omega s\right]}{\sqrt{\cosh \Xi_n^\mp - \cosh\left[\frac{r_h}{\ell^2} s\right]}}. \tag{5.48}$$

Using Eq. (5.48), X simplifies to

$$X = -\lambda^2 K \sum_{n=-\infty}^\infty \int_0^\infty dy \, e^{-\gamma y^2} \cos\theta y$$

$$\times \left[\frac{1}{\sqrt{\cosh \Xi_n^- - \cosh y}} - \zeta \frac{1}{\sqrt{\cosh \Xi_n^+ - \cosh y}} \right], \tag{5.49}$$

where we have changed the integration variable to $y := r_h s/\ell^2$ and defined

$$K := \sqrt{\frac{b_A b_B}{b_A^2 + b_B^2}} \frac{\sigma}{2\sqrt{\pi}} e^{-\frac{(b_A + b_B)^2}{b_A^2 + b_B^2}\frac{\Omega^2 \sigma^2}{2}}, \tag{5.50a}$$

$$\gamma := \frac{b_A^2 b_B^2}{b_A^2 + b_B^2} \frac{\ell^4}{r_h^2} \frac{1}{2\sigma^2}, \tag{5.50b}$$

$$\theta := \frac{b_A b_B}{b_A^2 + b_B^2}(b_B - b_A)\frac{\ell^2}{r_h}\Omega. \tag{5.50c}$$

As expressed in Eq. (5.49), X may now be evaluated numerically.

Having brought both P_D and X into a form that can be evaluated numerically,[11] we may compute the concurrence $\mathcal{C}(\rho_{AB})$ using Eq. (5.38). The concurrence is plotted as a function of the proper distance[12] the detectors are away from the BTZ horizon in Fig. 5.5 and as a function of the proper distance between the detectors in Fig. 5.6. In both figures $\sigma = 1$, $M = 1$, $\ell/\sigma = 10$, and the field satisfies Dirichlet boundary conditions ($\zeta = 1$) at spatial infinity.

In Fig. 5.5 we see that detectors placed closer to the horizon become less entangled, and at a finite distance away from the horizon do not become entangled at all. To interpret this behaviour it is helpful to examine the concurrence $\mathcal{C}(\rho_{AB})$ given in Eq. (5.38). We see that the larger the difference between $|X|$ and $\sqrt{P_A P_B}$, the greater the entanglement is in the final state of the two detectors; when $\sqrt{P_A P_B} \geq |X|$ the detectors do not become entangled. As the detectors move closer to the horizon, they experience a greater Hawking temperature and thus their probability of transitioning to their excited state is greater, increasing $\sqrt{P_A P_B}$. However, the dominate effect governing the difference between $|X|$ and $\sqrt{P_A P_B}$ is the decrease in $|X|$ as the detectors move closer to the horizon. This is a result of the red shift factors b_A and b_B appearing in the expression for X given in Eq. (5.39c), which vanish as the detectors approach the horizon.

In Fig. 5.5 we also observe that detectors with larger energy gaps are able to become entangled closer to the horizon. This is due to the fact that detectors with a larger energy gap are harder to excite, which results in the term $\sqrt{P_A P_B}$ being smaller for such detectors.

In Fig. 5.6 we observe that as the separation between the detectors grows, the entanglement between the detectors decreases. This is because correlations in the vacuum state are small for spacetime points separated by a large distance, which can be seen from the BTZ Wightman function in Eq. (5.18). We also observe that the entanglement decreases more slowly for detectors with larger energy gap and vanishes for finite detector separation.

---

[11]The numerical calculations were carried out in Mathematica. The integrals appearing in Eqs. (5.40) and (5.49) were evaluated using NIntegrate with MaxRecursion → 40, WorkingPrecision → 15, and Method → "DoubleExponential". To generate the plots in Figs. 5.5 and 5.6, the P_D and X were evaluated for 100 points uniformly distributed across the domain of each plot.

[12]The proper distance between two points $x_1 = (t, R_1, \phi)$ and $x_2 = (t, R_2, \phi)$, where $R_2 \geq R_1 \geq r_h$, is

$$d(R_1, R_2) := \int_{R_1}^{R_2} dr \frac{\ell}{\sqrt{r^2 - r_h^2}} = \ell \ln \left[\frac{R_2 + \sqrt{R_2^2 - r_h^2}}{R_1 + \sqrt{R_1^2 - r_h^2}} \right].$$

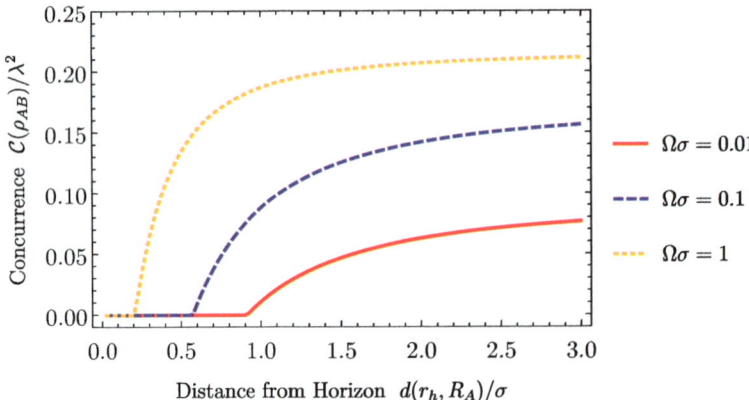

Fig. 5.5 The concurrence $\mathcal{C}(\rho_{AB})$ of the final state of two static detectors operating in the exterior region of the BTZ black hole is plotted as a function of the proper distance detector A is from the horizon. The proper distance between the detectors is set to $d(R_A, R_B)/\sigma = 1$, and $\ell/\sigma = 10$, $M = 1$, and the field satisfies Dirichlet boundary conditions ($\zeta = 1$)

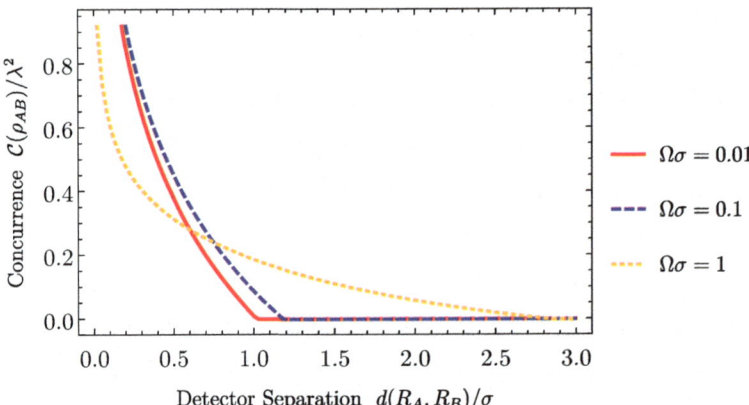

Fig. 5.6 The concurrence $\mathcal{C}(\rho_{AB})$ of the final state of two static detectors operating in the exterior region of the BTZ black hole is plotted as a function of the proper distance separating them. The proper distance between detector A and the horizon is $d(r_h, R_A)/\sigma = 1$, and $\ell/\sigma = 10$, $M = 1$, and the field satisfies Dirichlet boundary conditions ($\zeta = 1$)

5.5 Summary

We began in Sect. 5.1 by constructing both the BTZ and $\mathbb{R}\mathbf{P}^2$ geon spacetimes from AdS$_3$ by appropriate topological identifications. In Sect. 5.2, using the fact that both the BTZ and $\mathbb{R}\mathbf{P}^2$ geon black holes are quotient spacetimes, we constructed the Hartle-Hawking vacuum Wightman function associated with a conformally coupled massless scalar field on both the BTZ and $\mathbb{R}\mathbf{P}^2$ geon spacetimes from the AdS$_3$ Wightman functions via the method of images.

In Sect. 5.3 we used these Wightman functions to compute the transition rate of an Unruh-DeWitt detector operating in the exterior region of the BTZ and $\mathbb{R}\mathbf{P}^2$ geon black holes. We saw that even though the metric in both spacetimes is identical in the region where the detectors were operating, a difference in their transition rates was observed. The transition rate of a detector in the $\mathbb{R}\mathbf{P}^2$ geon spacetime turned on in the asymptotic past will develop a time dependence in the future of the preferred time hypersurface $t = 0$. Specifically, it will oscillate around the transition rate of an identical detector in the BTZ spacetime, which is time-independent. We conclude, in principle, a detector operating in the exterior region of the $\mathbb{R}\mathbf{P}^2$ geon black hole is sensitive to the non-stationary features of the spacetime located behind the past and future horizons. Information about the global structure of the $\mathbb{R}\mathbf{P}^2$ geon spacetime is encoded in the vacuum state of the field, and in principle accessible by local measurements of the field with Unruh-DeWitt detectors. Since the publication of these results [24], similar effects have been observed for detectors in the Schwarzschild spacetime and the related $\mathbb{R}\mathbf{P}^3$ geon spacetime [23].

In Sect. 5.4 we applied the entanglement harvesting protocol developed in Chap. 3 to detectors operating in the exterior region of the BTZ black hole. We derived the relevant matrix elements (P_A, P_B, and X) of the final state ρ_{AB} of the two detectors necessary to compute the concurrence $\mathcal{C}(\rho_{AB})$. The dependence of this entanglement on the detector separation and proximity of the detectors to the BTZ horizon was explained in terms of the response of the detectors to the local Hawking temperature and red shift effects.

The purpose of investigating the entanglement harvesting protocol for detectors in the BTZ spacetime is twofold: it serves as a concrete application of the formalism developed in Chap. 3 and begins an investigation into how the entanglement structure of a quantum field theory depends on the underlying spacetime geometry as seen by local measurements of the field. The hope is that the entanglement harvesting protocol can be applied in other spacetime geometries to better understand the connection between the entanglement structure of a quantum field theory and the properties of a spacetime on which the field is defined.

References

1. S.J. Avis, C.J. Isham, D. Storey, Quantum field theory in anti-de sitter spacetime. Phys. Rev. D **18**, 3565 (1978)
2. M. Bañados, C. Teitelboim, J. Zanelli, The black hole in three-dimensional spacetime. Phys. Rev. Lett. **69**, 1849 (1992)
3. M. Bañados, M. Henneaux, C. Teitelboim, J. Zanelli, Geometry of the 2+1 black hole. Phys. Rev. D **48**, 1506 (1993)
4. N.D. Birrell, P.C.W. Davies, *Quantum Fields in Curved Space* (Cambridge University Press, Cambridge, 1982)
5. S. Carlip, *Quantum Gravity in 2+1 Dimensions* (Cambridge University Press, Cambridge, 2003)
6. C.J. Fewster, B.A. Juárez-Aubry, J. Louko, Waiting for Unruh. Classical Quantum Gravity **33**, 165003 (2016)

7. L.J. Garay, E. Martín-Martínez, J. de Ramon, Thermalization of particle detectors: the Unruh effect and its reverse. Phys. Rev. D **94**, 104048 (2016)
8. M. Guica, S.F. Ross, Behind the geon horizon. Classical Quantum Gravity **32**, 055014 (2015)
9. L.J. Henderson, R.A. Hennigar, R.B. Mann, A.R.H. Smith, J. Zhang, Harvesting entanglement from the black hole vacuum. Classical Quantum Gravity **35**, 21LT02 (2018)
10. L. Hodgkinson, J. Louko, Static, stationary and inertial Unruh-DeWitt detectors on the BTZ black hole. Phys. Rev. D **86**, 064031 (2012)
11. L. Hodgkinson, J. Louko, Unruh-DeWitt detector on the BTZ black hole (2012). arXiv:gr-qc/1208.3165
12. C.J. Isham, Prima facie questions in quantum gravity, in *Canonical Gravity: From Classical to Quantum*, ed. by J. Ehlers, H. Friedrich (Springer, Berlin, 1994), pp. 1–21
13. R. Kubo, Statistical-mechanical theory of irreversible processes. I. General theory and simple applications to magnetic and conduction problems. J. Phys. Soc. Jpn. **12**, 570 (1957)
14. G. Lifschytz, M. Ortiz, Scalar field quantization on the 2+1 dimensional black hole background. Phys. Rev. D **49**, 1929 (1994)
15. J. Louko, Single exterior black holes, in *Towards Quantum Gravity*. Lecture Notes in Physics, vol. 541 (Springer, Berlin, 2000), pp. 188–202
16. J. Louko, Geon black holes and quantum field theory. J. Phys. Conf. Ser. **222**, 012038 (2010)
17. J. Louko, D. Marolf, Inextendible Schwarzschild black hole with a single exterior: how thermal is the Hawking radiation? Phys. Rev. D **58**, 024007 (1998)
18. J. Louko, D. Marolf, Single exterior black holes and the AdS/CFT conjecture. Phys. Rev. D **59**, 066002 (1999)
19. J. Louko, R.B. Mann, D. Marolf, Geons with spin and charge. Classical Quantum Gravity **22**, 1451 (2005)
20. P.C. Martin, J. Schwinger, Theory of many-particle systems. Phys. Rev. **115**, 1342 (1959)
21. C.W. Misner, J.A. Wheeler, Classical physics as geometry: gravitation, electromagnetism, unquantized charge, and mass as properties of curved empty space. Ann. Phys. **2**, 525 (1957)
22. V.F. Mukhanov, S. Winitzki, *Introduction to Quantum Effects in Gravity* (Cambridge University Press, Cambridge, 2007)
23. K. Ng, R.B. Mann, Over the horizon: distinguishing the Schwarzschild spacetime and the $\mathbb{RP}^3$ spacetime using an Unruh-DeWitt detector. Phys. Rev. D **96**, 085004 (2017)
24. A.R.H. Smith, R.B. Mann, Looking inside a black hole. Classical Quantum Gravity **31**, 082001 (2014)
25. R.D. Sorkin, Introduction to topological geons, in *Topological Properties and Global Structure of Space-Time*, ed. by P.G. Bergmann, V.D. Sabbata. NATO ASI Series (Plenum, New York, 1986), pp. 249–270
26. R.M. Wald, *General Relativity* (The University of Chicago Press, Chicago, 1984)
27. J.A. Wheeler, Geons. Phys. Rev. **97**, 511 (1955)

Part II
Quantum Reference Frames

Chapter 6
Quantum Reference Frames Associated with Noncompact Groups

When we describe the configuration of a system, we almost always make use of a classical reference frame. Suppose we wish to specify the speed of a boat traveling along a flowing river. To do so we need a reference frame, which we usually take to be either the moving water or the river bank. Everyday when we define up and down, we use as a reference frame the gravitational field of the Earth. Indeed, the lesson of relativity is that physical quantities only have meaning with respect to a reference frame.

The same is true in quantum theory. When defining a basis for a Hilbert space we employ a classical reference frame. For example, when defining the quantization axis of a spin system we may make use of a classical magnetic field in a Stern-Gerlach device. In the study of quantum fields on curved spacetime, the reference frame with respect to which the fields are defined is the spacetime itself, as seen by an observer employing a suitable coordinate system.

This state of affairs is not fully satisfactory for one notable reason: a quantum system is being described with respect to a classical system, mixing elements from conceptually different frameworks. We must remember that a reference frame is a physical object, and as such it too is subject to the laws of quantum mechanics. At some scale, the quantum properties of a reference frame will affect our description of a system. Therefore, we need to ask: What happens when we replace a classical reference frame with a quantum one?

The first to explore this question were Aharonov and Susskind [7, 8] who showed that superselection rules may be lifted using an appropriate reference frame, a point that has since been emphasized by many authors [15, 16, 20, 24, 25]. Shortly after, quantum aspects of the equivalence principle were studied [5] and it was demonstrated that quantum reference frames can be consistently incorporated in quantum theory [6]. Since then, the study of quantum reference frames has taken on an increasingly information-theoretic flavour [12], finding practical applications in quantum interferometry [17], quantum communication [13] (see Chap. 7), cryptography [19], and resource theories [23].

© Springer Nature Switzerland AG 2019
A. R. H. Smith, *Detectors, Reference Frames, and Time*, Springer Theses,
https://doi.org/10.1007/978-3-030-11000-0_6

Quantum reference frames have also proven useful in the study of quantum gravity. General relativity does not make use of a reference frame in its construction; it is a background independent theory[1] and there is no a priori reason why its quantization should introduce a reference frame. Thus, we might reasonably expect that a quantum theory of gravity will be background independent. These considerations have led to the aspiration of building a relational quantum mechanics [30, 31]. Poulin has constructed a relational formulation of quantum theory by making explicit use of quantum reference frames [27]. Furthermore, Rovelli has shown that constructing physical observables in a generally covariant theory requires the inclusion of the dynamics of the objects serving as reference frames [29], and studied the consequences of this in quantum theory [28].

The natural language with which to describe reference frames is group theory, owing to the fact that the transformations that describe the act of changing reference frames form a group. Most discussion of quantum reference frames revolves around reference frames defined with respect to compact groups. For example, the group used to describe a change of phase reference in quantum optics is $U(1)$ and the group used to describe the transformation between orientations of a laboratory is $SO(3)$.

However, there are reference frames associated with noncompact groups that are of physical interest. For example, special relativity is essentially the study of reference frames associated with the Poincaré group. To study the quantum properties of reference frames associated with these groups, the existing formalism used to study quantum reference frames associated with compact groups will need to be generalized to noncompact groups. The purpose of this chapter and the one that follows is to embark on such a task by studying reference frames associated with the noncompact group of translations and the noncompact group of Galilean boosts.

We begin in Sect. 6.1 by introducing the G-twirl, which is a group average over all possible orientations of a system with respect to an external reference frame, which may be used to construct a relational description of a quantum system with respect to a quantum reference frame. We then demonstrate the failure of the G-twirl and this relational description when naively applied to situations involving the noncompact groups of translations and Galilean boosts. However, we find that the G-twirl over these groups naturally introduces a reduced state obtained by tracing out the centre-of-mass degrees of freedom of a composite system. In Sect. 6.2 we examine informational properties of this reduced state for systems of two and three particles in fully separable Gaussian states with respect to an external frame. Specifically, we study the entanglement that appears when moving from a description of the system with respect to an external frame to a fully relational description, which can alternatively be interpreted in terms of noise. This study is motivated by the need to determine how best to prepare states in the external

---

[1]This is not quite true. Background structure such as topology, spacetime dimension, and metric signature still exist, and may or may not be subject to quantization. See [33] for further discussion.

partition in order to encode information in relational degrees of freedom, which will be useful for various communication tasks [14]. We conclude this chapter in Sect. 6.3 with a discussion and summary of the results presented.

6.1 Relational Descriptions

In the construction of a relational quantum theory, an essential task is the description of a quantum system with respect to another quantum system. With this in mind, we seek a way in which to remove any information contained in a quantum state about its relation to an external reference frame. This is accomplished by the G-twirl, which we introduce in Sect. 6.1.1 and apply to the group of translations and Galilean boosts in Sect. 6.1.2.

6.1.1 Relational Description for Compact Groups

When the state of a system is described with respect to an external reference frame, such that changes of this reference frame form a compact group, a relational description constructed using the G-twirl is well studied [12].

Suppose we have a quantum system in the state $\rho \in S(\mathcal{H})$, where $\mathcal{H}$ is the Hilbert space associated with the system, described with respect to an external reference frame. Changes of the orientation of the system with respect to the external frame are implemented by $U(g) \in \mathcal{U}(\mathcal{H})$ acting on ρ, where $U(g)$ is the unitary representation of the group element $g \in G$, and G is the compact group of all possible changes of the external reference frame. The relational description of ρ, that is, the quantum state that does not contain any information about the external frame, is given by an average over all possible orientations of ρ with respect to the external frame, with each possible orientation given an equal weight

$$\mathcal{G}[\rho] := \int dg \, U(g) \, \rho \, U^{\dagger}(g), \tag{6.1}$$

where dg is the Haar measure associated with the group G. This averaging operation is known as the G-twirl. For compact groups the G-twirl takes states within the state space $S(\mathcal{H})$ to other states in $S(\mathcal{H})$

$$\mathcal{G} : S(\mathcal{H}) \to S(\mathcal{H}). \tag{6.2}$$

By averaging over all elements of the group, the G-twirl removes any relation to the external reference frame that was implicitly made use of in the description of ρ. What remains is only information about the relational degrees of freedom within the system, that is, information unaffected by changes of the external reference frame.

For example, if $\rho \in S(\mathcal{H})$ describes a composite system of two particles such that $\mathcal{H} = \mathcal{H}_1 \otimes \mathcal{H}_2$, what remains in $\mathcal{G}[\rho]$ is information about the relational degrees of freedom between the two particles. Note that the G-twirl is performed via the product representation $U(g) = U_1(g) \otimes U_2(g)$, where $U_1(g) \in \mathcal{U}(\mathcal{H}_1)$ and $U_2(g) \in \mathcal{U}(\mathcal{H}_2)$ are representations of the group G on each of the subsystems.

This relational description is used extensively in the study of quantum reference frames involving compact groups [12, 13, 17, 22, 26]. However, when the G-twirl operation is generalized to the case where the group G is noncompact, and thus does not admit a normalized Haar measure, it results in non-normalizable states.

To illustrate this point consider the G-twirl of the state $\rho \in S(\mathcal{H})$, where $\mathcal{H} \simeq L_2(\mathbb{R})$, over the noncompact group of spatial translations T_1 generated by the momentum operator P. Expressing ρ in the momentum basis, we find

$$\mathcal{G}_{T_1}[\rho] = \int dg \, e^{-igP} \left(\int dp dp' \, \rho(p, p') \, |p\rangle\langle p'| \right) e^{igP}$$

$$= 2\pi \int dp \, \rho(p, p) \, |p\rangle\langle p| \, , \tag{6.3}$$

where dg is the Haar measure associated with T_1 and in going from the first to the second equality we have used the definition of the Dirac delta function $2\pi\delta(p - p') := \int dg \, e^{ig(p-p')}$. Although the averaging operation is mathematically well defined, the resulting state $\mathcal{G}[\rho]$ is not normalized, as the trace of $\mathcal{G}_{T_1}[\rho]$ is infinite.

For noncompact groups, the action of the G-twirl maps states to a space outside of the state space $S(\mathcal{H})$. This is a consequence of the Haar measure not being normalized, i.e. the integral $\int dg$ is infinite for noncompact groups. This issue does not arise when twirling over a compact group for which there exists a normalized Haar measure. Thus, the relational description constructed by averaging a system over all possible orientations of a reference frame fails when the group describing changes of the reference frame is noncompact.

One may try to remedy this problem by introducing a measure $p(g)$ on the group such that $\int dg \, p(g) = 1$, and interpreting $p(g)$ as representing a priori knowledge of how the average should be performed [9]. However, in general there is no objective way to choose $p(g)$—if we want a normalized measure it cannot be invariant.

6.1.2 Relational Description for Noncompact Groups

We now construct a relational description of quantum states suitable for systems described with respect to reference frames associated with the noncompact groups of boosts and translations. We begin by twirling the state of a system of particles $\rho \in S(\mathcal{H})$ over all possible boosts and translations of the external reference frame with respect to which ρ is defined. The result of this twirling is a non-normalizable state proportional to $I_{CM} \otimes \rho_R$, where I_{CM} is the identity on the centre-of-mass degrees of freedom and $\rho_R := \mathrm{tr}_{CM}\, \rho$ is a normalized density matrix describing the relative degrees of freedom of the system. In doing so, we connect two approaches to quantum reference frames that have been studied in the past: the approach introduced by Bartlett et al. [12], which makes use of the G-twirl to remove any information the state may have about an external reference frame, and the approach of Angelo et al. [11], in which a partial trace over centre-of-mass degrees of freedom is used to obtain a relational state.

Consider a composite system of N particles each with mass m_n. We may partition the Hilbert space $\mathcal{H}$ of the entire system as $\mathcal{H} = \bigotimes_{n=1}^{N} \mathcal{H}_n$ where $\mathcal{H}_n \simeq L_2(\mathbb{R}^3)$, which spans the degrees of freedom defined with respect to an external frame associated with the nth particle; we will refer to this as the external partition of the Hilbert space. We may alternatively partition the Hilbert space as $\mathcal{H} = \mathcal{H}_{CM} \otimes \mathcal{H}_R$, where $\mathcal{H}_{CM} \simeq L_2(\mathbb{R}^3)$ is associated with the degrees of freedom of the centre-of-mass defined with respect to an external frame, and $\mathcal{H}_R \simeq L_2(\mathbb{R}^{3N-3})$ is associated with the relative degrees of freedom of the system defined with respect to a chosen reference particle; we will refer to this partition as the centre-of-mass and relational partition of the Hilbert space.

As was done in Sect. 6.1.1 for reference frames associated with compact groups, to obtain a relational state we will average the state of our system over all possible orientations—intended in a generic sense, meant here to be about translations and boosts—with respect to the external frame. Here we consider the system to be described with respect to an inertial external frame. Thus a change of the external frame corresponds to acting on the system with an element of the Galilean group, and the average over all possible orientations of the system with respect to the external frame will be an average over the Galilean group.

The Galilean group Gal is a semidirect product of the translation group T_4, the group of boosts B_3, and the rotation group $SO(3)$,

$$Gal \simeq T_4 \rtimes \Big(B_3 \rtimes SO(3)\Big). \tag{6.4}$$

We will limit our analysis to averaging over spatial translations T_3, where $T_4 \simeq T_1 \rtimes T_3$, and boosts B_3. We will not average over rotations $SO(3)$, since this has been well studied elsewhere [12] and we are primarily interested in issues associated with noncompact groups. Further, we do not average over time translations T_1, as this would require us to introduce a Hamiltonian to generate time translations, and for now we are interested only in a relative description of the state at one instant of time and not its dynamics. Suppose the state of a system is given with respect to an external reference frame with a specific position and velocity. The operator that results from these restricted averages is the state as seen by an observer who is ignorant of both the position and velocity of the external reference frame.

The position and momentum operators associated with the centre-of-mass, $\mathbf{X}_{CM}$ and $\mathbf{P}_{CM}$, and relational degrees of freedom, $\mathbf{X}_{i|1}$ and $\mathbf{P}_{i|1}$, may be expressed in terms of the operators $\mathbf{X}_n$ and $\mathbf{P}_n$ associated with the position and momentum operators of each of the N particles with respect to the external frame as

$$\mathbf{X}_{CM} = \frac{1}{M} \sum_{n=1}^{N} m_n \mathbf{X}_n, \tag{6.5a}$$

$$\mathbf{P}_{CM} = \sum_{n=1}^{N} \mathbf{P}_n, \tag{6.5b}$$

$$\mathbf{X}_{i|1} = \mathbf{X}_i - \mathbf{X}_1 \ \text{ for } \ i \in \{2, \dots, N\}, \tag{6.5c}$$

$$\mathbf{P}_{i|1} = \mathbf{P}_i - \frac{m_i}{M} \mathbf{P}_{CM} \ \text{ for } \ i \in \{2, \dots, N\}, \tag{6.5d}$$

where $M := \sum_{n=1}^{N} m_n$ is the total mass, and without loss of generality we have chosen to define the relative degrees of freedom with respect to particle 1. The above operators satisfy the canonical commutation relations $[\mathbf{X}_{CM}, \mathbf{P}_{CM}] = [\mathbf{X}_{i|1}, \mathbf{P}_{i|1}] = i$, with all other combinations vanishing. Defining the relative degrees of freedom in this way specifies particle 1 as the reference frame for the relational degrees of freedom associated with the other particles. This will allow us to associate properties of particle 1, such as its mass and quantum state, with properties of the reference frame used in the relational description given below.

With the exception of the two-particle case, $\mathbf{P}_{i|1}$ is not equal to the usually defined relative momentum operator

$$\mathbf{P}_{r_i} := \mu_{1i} \left(\frac{\mathbf{P}_i}{m_i} - \frac{\mathbf{P}_1}{m_1} \right) \neq \mathbf{P}_{i|1} \ \text{ for } \ i \in \{2, \dots, N\}, \tag{6.6}$$

where $\mu_{1i} := m_1 m_i / (m_1 + m_i)$ is the reduced mass of particle 1 and the ith particle, as one might expect. Alternatively, one may begin with the set of relative momentum operators $\{\mathbf{P}_{r_i} \,|\, i = 2, \dots, N\}$ and construct canonically conjugate relative position operators. However, we restrict ourselves to considering the operators given in Eq. (6.5) and refer the reader to [11] for a more detailed discussion of the nonuniqueness of canonically conjugate operators on $\mathcal{H}_R$.

The action of a translation $\mathbf{g} \in \mathbb{R}^3 \simeq T_3$ and boost $\mathbf{h} \in \mathbb{R}^3 \simeq B_3$ of the external frame in the external partition $\mathcal{H} = \bigotimes_{n=1}^{N} \mathcal{H}_n$ is given by

$$U_T(\mathbf{g}) = \bigotimes_{n=1}^{N} e^{-i\mathbf{g}\cdot\mathbf{P}_n}, \tag{6.7a}$$

$$U_B(\mathbf{h}) = \bigotimes_{n=1}^{N} e^{im_n\mathbf{h}\cdot\mathbf{X}_n}, \tag{6.7b}$$

and in the centre-of-mass and relational partition $\mathcal{H}_{CM} \otimes \mathcal{H}_R$ is given by

$$U_T(\mathbf{g}) = e^{-i\mathbf{g}\cdot\mathbf{P}_{CM}} \otimes I_R, \tag{6.8a}$$

$$U_B(\mathbf{h}) = e^{iM\mathbf{h}\cdot\mathbf{X}_{CM}} \otimes I_R. \tag{6.8b}$$

To carry out the average over T_3 and B_3, let us express ρ in the $\mathcal{H}_{CM} \otimes \mathcal{H}_R$ partition in the momentum basis

$$\rho = \int d\mathbf{p}_{CM} d\mathbf{p}'_{CM} d\mathbf{p}_R d\mathbf{p}'_R \, \rho(\mathbf{p}_{CM}, \mathbf{p}'_{CM}, \mathbf{p}_R, \mathbf{p}'_R) \, |\mathbf{p}_{CM}\rangle\langle\mathbf{p}'_{CM}| \otimes |\mathbf{p}_R\rangle\langle\mathbf{p}'_R|,$$
(6.9)

where $\mathbf{p}_{CM}$ and $\mathbf{p}'_{CM}$ denote the momentum vector of the centre-of-mass and $\mathbf{p}_R$ and $\mathbf{p}'_R$ denote the $N - 1$ relative momentum vectors. Making use of Eq. (6.8a), we may average over all possible spatial translations of the external frame

$$\mathcal{G}_T[\rho] = \int d\mathbf{p}_{CM} d\mathbf{p}'_{CM} d\mathbf{p}_R d\mathbf{p}'_R \, \rho(\mathbf{p}_{CM}, \mathbf{p}'_{CM}, \mathbf{p}_R, \mathbf{p}'_R)$$
$$\times \int d\mathbf{g} \, U_T(\mathbf{g}) \Big[|\mathbf{p}_{CM}\rangle\langle\mathbf{p}'_{CM}| \otimes |\mathbf{p}_R\rangle\langle\mathbf{p}'_R| \Big] U_T(\mathbf{g})^\dagger$$
$$= (2\pi)^3 \int d\mathbf{p}_{CM} d\mathbf{p}_R d\mathbf{p}'_R \, \rho(\mathbf{p}_{CM}, \mathbf{p}_{CM}, \mathbf{p}_R, \mathbf{p}'_R) \, |\mathbf{p}_{CM}\rangle\langle\mathbf{p}_{CM}| \otimes |\mathbf{p}_R\rangle\langle\mathbf{p}'_R|.$$
(6.10)

From Eq. (6.10) we see that the effect of twirling over the group of translations T_3 is to project ρ into a charge sector of definite centre-of-mass momentum. That is, lacking a reference frame associated with the translation group imposes a superselection rule forbidding coherence between different centre-of-mass momentum eigenstates.

Similarly, we can average ρ over all possible boosts of the external frame with the result

$$\mathcal{G}_B[\rho] = \int d\mathbf{x}_{CM} d\mathbf{x}'_{CM} d\mathbf{x}_R d\mathbf{x}'_R \, \rho(\mathbf{x}_{CM}, \mathbf{x}'_{CM}, \mathbf{x}_R, \mathbf{x}'_R)$$
$$\times \int d\mathbf{h} \, U_B(\mathbf{h}) \Big[|\mathbf{x}_{CM}\rangle\langle\mathbf{x}'_{CM}| \otimes |\mathbf{x}_R\rangle\langle\mathbf{x}'_R| \Big] U_B(\mathbf{h})^\dagger$$
$$= \left(\frac{2\pi}{M}\right)^3 \int d\mathbf{x}_{CM} d\mathbf{x}_R d\mathbf{x}'_R \, \rho(\mathbf{x}_{CM}, \mathbf{x}_{CM}, \mathbf{x}_R, \mathbf{x}'_R) \, |\mathbf{x}_{CM}\rangle\langle\mathbf{x}_{CM}| \otimes |\mathbf{x}_R\rangle\langle\mathbf{x}'_R|,$$
(6.11)

where $\mathbf{x}_{CM}$ and $\mathbf{x}'_{CM}$ denote the position vector of the centre-of-mass, $\mathbf{x}_R$ and $\mathbf{x}'_R$ denote the $N - 1$ relative position vectors, and $\rho(\mathbf{x}_{CM}, \mathbf{x}'_{CM}, \mathbf{x}_R, \mathbf{x}'_R) = \langle\mathbf{x}_{CM}| \langle\mathbf{x}_R| \rho |\mathbf{x}_{CM}\rangle |\mathbf{x}_R\rangle$. From Eq. (6.11) we see the effect of twirling over the group of boosts B_3 is to project ρ into a charge sector of definite centre-of-mass position. That is, lacking a reference frame associated with the group of boosts imposes a superselection rule forbidding coherence between different centre-of-mass position eigenstates.

Averaging Eq. (6.10) over all boosts, using Eq. (6.8b), yields

$$\mathcal{G}_B \circ \mathcal{G}_T[\rho] = (2\pi)^3 \int d\mathbf{h} \int d\mathbf{p}_{CM} d\mathbf{p}_R d\mathbf{p}'_R \, \rho(\mathbf{p}_{CM}, \mathbf{p}_{CM}, \mathbf{p}_R, \mathbf{p}'_R)$$

$$\times U_B(\mathbf{h}) \big[|\mathbf{p}_{CM}\rangle\langle\mathbf{p}_{CM}| \otimes |\mathbf{p}_R\rangle\langle\mathbf{p}'_R| \big] U_B(\mathbf{h})^\dagger$$

$$= (2\pi)^3 \int d\mathbf{h} \int d\mathbf{p}_{CM} d\mathbf{p}_R d\mathbf{p}'_R \, \rho(\mathbf{p}_{CM} - M\mathbf{h}, \mathbf{p}_{CM} - M\mathbf{h}, \mathbf{p}_R, \mathbf{p}'_R)$$

$$\times |\mathbf{p}_{CM}\rangle\langle\mathbf{p}_{CM}| \otimes |\mathbf{p}_R\rangle\langle\mathbf{p}'_R|$$

$$= \left(\frac{2\pi}{M}\right)^3 \int d\mathbf{h} \int d\mathbf{p}_{CM} d\mathbf{p}_R d\mathbf{p}'_R \, \rho(\mathbf{h}, \mathbf{h}, \mathbf{p}_R, \mathbf{p}'_R) \, |\mathbf{p}_{CM}\rangle\langle\mathbf{p}_{CM}|$$

$$\otimes |\mathbf{p}_R\rangle\langle\mathbf{p}'_R|$$

$$= \left(\frac{2\pi}{M}\right)^3 \int d\mathbf{p}_{CM} \, |\mathbf{p}_{CM}\rangle\langle\mathbf{p}_{CM}|$$

$$\otimes \int d\mathbf{p}_R d\mathbf{p}'_R \left(\int d\mathbf{h} \, \rho(\mathbf{h}, \mathbf{h}, \mathbf{p}_R, \mathbf{p}'_R) \right) |\mathbf{p}_R\rangle\langle\mathbf{p}'_R|$$

$$= \left(\frac{2\pi}{M}\right)^3 I_{CM} \otimes \rho_R, \tag{6.12}$$

where in the last line

$$\rho_R := \int d\mathbf{p}_R d\mathbf{p}'_R \left(\int d\mathbf{h} \, \rho(\mathbf{h}, \mathbf{h}, \mathbf{p}_R, \mathbf{p}'_R) \right) |\mathbf{p}_R\rangle\langle\mathbf{p}'_R| = \mathrm{tr}_{CM} \, \rho, \tag{6.13}$$

and we have made use of the resolution of the identity $I_{CM} = \int d\mathbf{p}_{CM} \, |\mathbf{p}_{CM}\rangle\langle\mathbf{p}_{CM}|$.
The action of $\mathcal{G}_B \circ \mathcal{G}_T$ may be expressed as

$$\mathcal{G}_B \circ \mathcal{G}_T[\rho] = \left(\frac{2\pi}{M}\right)^3 \left(\mathcal{D}_{CM} \otimes \mathcal{I}_R \right) [\rho], \tag{6.14}$$

where $\mathcal{D}_{CM}$ denotes the operation that takes every operator on $\mathcal{H}_{CM}$ to the identity
operator on that space and $\mathcal{I}_R$ denotes the identity map on $\mathcal{H}_R$. Note that the
generators of T_3 and B_3 commute to a multiple of the identity, $[\mathbf{X}_{CM}, \mathbf{P}_{CM}] = i I_{CM}$, and consequently by application of the Baker-Campbell-Hausdorff equality
it can be shown that $\mathcal{G}_B \circ \mathcal{G}_T = \mathcal{G}_T \circ \mathcal{G}_B$. From the appearance of $\mathcal{D}_{CM}$, the
analogue of the completely depolarizing channel on $\mathcal{H}_{CM} \simeq L_2(\mathbb{R}^3)$, in Eq. (6.14),
we see that $\mathcal{G}_B \circ \mathcal{G}_T[\rho]$ contains no information about the centre-of-mass, and thus
no information about the external frame. However, all the information about the
relational degrees of freedom of the system is encoded in ρ_R, which is normalized.

By twirling over all possible boosts and translations of the system, we see from
Eq. (6.12) that the reduced state ρ_R naturally appears. This demonstrates how the
relational state ρ_R, which is used by Angelo et al. [10, 11] when analysing absolute

and relative degrees of freedom, is obtained from the usual quantum reference formalism [12].

Summarizing, the relational state is given as the output of a map Λ acting on $\rho \in \mathcal{S}(\mathcal{H})$:

$$\Lambda : \mathcal{S}(\mathcal{H}) \rightarrow \mathcal{S}(\mathcal{H}_R),$$

$$\rho \mapsto \rho_R = \mathrm{tr}_{CM}\, \rho. \tag{6.15}$$

The map Λ is qualitatively different than the G-twirl, as the domain of Λ is not equal to its range.

In general, when transforming from the external partition $\mathcal{H} = \bigotimes_{n=1}^{N} \mathcal{H}_n$, to the centre-of-mass and relational partition $\mathcal{H} = \mathcal{H}_{CM} \otimes \mathcal{H}_R$, entanglement will appear between the centre-of-mass and relational degrees of freedom, as well as within the relational Hilbert space $\mathcal{H}_R$. As a consequence the state ρ_R will be mixed, reflecting the fact that information about the external degrees of freedom has been lost. This is analogous to information about the external frame being lost in Eq. (6.1) when averaging over all elements of a compact group. In the next section we will quantify this information loss for systems of two and three particles in Gaussian states.

6.2 Gaussian Quantum Reference Frames

We now examine in detail the informational properties of the reduced state ρ_R of the relational degrees of freedom given in Eq. (6.13) by examining systems of two and three particles in one dimension distinguished by their masses. As mentioned in Sect. 6.1, in general, entanglement will appear when moving from the external partition $\mathcal{H} = \bigotimes_{n=1}^{N} \mathcal{H}_n$ to the centre-of-mass and relational partition $\mathcal{H} = \mathcal{H}_{CM} \otimes \mathcal{H}_R$. This entanglement is crucial in determining how to describe physics relative to a particle within the system [11]. For example, if there is entanglement between the centre-of-mass and the relational degrees of freedom, an observer identified with the reference particle, particle 1 as chosen in Eq. (6.5), will describe the rest of the system as being in a mixed state.

As a concrete example of the entanglement that can emerge when changing from the external partition to the centre-of-mass and relational partition of the Hilbert space, we consider systems of two and three particles in Gaussian states in the external partition. The advantage of considering Gaussian states in the external partition is that the transformation which takes the state from being specified in the external partition to being specified in the centre-of-mass and relational partition is a Gaussian unitary, that is, a state which is Gaussian in the external partition will also be Gaussian in the centre-of-mass and relational partition. Further, if we are interested in the reduced state ρ_R defined in Eq. (6.13), and the state of the particles in either partition is a Gaussian state, then the trace over the centre-of-mass degrees of freedom also results in a Gaussian state. Thus, by considering Gaussian states in the external partition we are able to make use of the extensive tools developed in the field of Gaussian quantum information. We begin here by briefly reviewing relevant aspects of Gaussian quantum information; for more detail the reader may consult one of the many good references on the topic [1, 3, 35].

6.2.1   The Wigner Function and Gaussian States

Any density operator has an equivalent representation as a quasiprobability distribution over phase space. To see this, we introduce the Weyl operator

$$D(\boldsymbol{\xi}) := \exp\!\left(i\mathbf{X}^T \boldsymbol{\Omega}\boldsymbol{\xi}\right),$$

(6.16)

where $\mathbf{X} := (Q_1, P_1, \ldots, Q_n, P_n)$ is a vector of phase space operators, $\boldsymbol{\xi} \in \mathbb{R}^{2n}$, and Ω is the symplectic form defined as

$$\Omega := \bigoplus_{i=1}^{n} \omega, \quad \text{where} \quad \omega := \begin{pmatrix} 0 & 1 \\ -1 & 0 \end{pmatrix}.$$

(6.17)

A density operator $\rho \in \mathcal{S}(\mathcal{H})$ has an equivalent representation as a Wigner characteristic function $\chi(\boldsymbol{\xi}) := \operatorname{tr}[\rho D(\boldsymbol{\xi})]$, or by its Fourier transform which is known as the Wigner function

$$W(\mathbf{x}) := \int_{\mathbb{R}^{2n}} \frac{d^{2n}\xi}{(2\pi)^{2n}} \exp\!\left(-i\mathbf{x}^T \boldsymbol{\Omega}\boldsymbol{\xi}\right) \chi(\boldsymbol{\xi}),$$

(6.18)

where $\mathbf{x} := (q_1, p_1, \ldots, q_n, p_n)$ is a vector of phase space variables.

An n-particle Gaussian state is a state whose Wigner function is Gaussian

$$W(\mathbf{x}; \bar{\mathbf{x}}, \mathbf{V}) = \frac{\exp\!\left(-\frac{1}{2}(\mathbf{x} - \bar{\mathbf{x}})^T \mathbf{V}^{-1}(\mathbf{x} - \bar{\mathbf{x}})\right)}{(2\pi)^n \sqrt{\det \mathbf{V}}},$$

(6.19)

where $\bar{\mathbf{x}} := (\bar{q}_1, \bar{p}_1, \ldots, \bar{q}_n, \bar{p}_n)$ is given by a vector of averages

$$\bar{x}_i := \langle X_i \rangle = \operatorname{tr}[X_i \rho],$$

(6.20)

and $\mathbf{V}$ is the real $2n \times 2n$ covariance matrix with components

$$V_{ij} := \frac{1}{2} \operatorname{tr}\left[\{X_i - \bar{x}_i, X_j - \bar{x}_j\} \rho\right],$$

(6.21)

where we have made use of the anticommutator $\{A, B\} := AB + BA$.

6.2.2   The Two-Particle Case

We begin our analysis by considering two particles with masses m_1 and m_2 to be in a tensor product of Gaussian states $\rho_E = \rho_1 \otimes \rho_2$ in the external partition $\mathcal{H} = \mathcal{H}_1 \otimes \mathcal{H}_2$, where $\rho_1 \in \mathcal{S}(\mathcal{H}_1)$ and $\rho_2 \in \mathcal{S}(\mathcal{H}_2)$. Due to the tensor product structure of ρ_E, the Wigner function of the composite system is a product of the Wigner functions associated with particles 1 and 2

$$W\left(\mathbf{x}; \bar{\mathbf{x}}_E, \mathbf{V}_E\right) = W\left(\mathbf{x}; \bar{\mathbf{x}}_1, \mathbf{V}_1\right) W\left(\mathbf{x}; \bar{\mathbf{x}}_2, \mathbf{V}_2\right). \tag{6.22}$$

The reason for considering factorized states in the external partition, apart from their common usage in the literature [13, 26], is that if we are to use the composite system for communication (see Chap. 7), the tensor product structure is easily prepared as it does not require an entangling operation.

As we will only be interested in the entanglement generated in moving from the external partition to the centre-of-mass and relational partition, we may, without loss of generality, set $\bar{\mathbf{x}}_1 = \bar{\mathbf{x}}_2 = 0$, as these averages can be arbitrarily adjusted via local unitary operations in either partition, and thus do not affect the entanglement properties under consideration.

Making use of Eq. (6.19), we find the covariance matrix associated with ρ_E is given by $\mathbf{V}_E = \mathbf{V}_1 \oplus \mathbf{V}_2$; the direct sum structure resulting from the fact that we chose ρ_E to be a factorized state with respect to the external partition. Using Williamson's theorem [36], one can show that the most general form of the covariance matrices $\mathbf{V}_1$ and $\mathbf{V}_2$ is

$$
\begin{aligned}
\mathbf{V}_i &= \frac{1}{\mu_i} \mathbf{R}\left(\theta_i\right) \mathbf{S}\left(2r_i\right) \mathbf{R}\left(\theta_i\right)^T \\
&= \frac{1}{\mu_i} \begin{pmatrix} \cosh 2r_i - \cos 2\theta_i \sinh 2r_i & \sin 2\theta_i \sinh 2r_i \\ \sin 2\theta_i \sinh 2r_i & \cosh 2r_i + \cos 2\theta_i \sinh 2r_i \end{pmatrix},
\end{aligned} \tag{6.23}
$$

where the free parameter $\mu_i = 1/\sqrt{\det \mathbf{V}_i} \in (0, 1]$ is the purity $\mathrm{tr}(\rho_i^2)$ of the state ρ_i, $\mathbf{R}\left(\theta_i\right)$ is a rotation matrix specifying a phase rotation by an angle $\theta_i \in [0, \pi/4]$, and $\mathbf{S}(2r_i)$ is a diagonal symplectic matrix specifying a squeezing of the Wigner function parameterized by $r_i \in \mathbb{R}$.

Transforming to the Centre-of-Mass and Relational Partition
For two particles in one dimension the transformation from the external degrees of freedom $\mathbf{x}_E := (x_1, p_1, x_2, p_2)$, where x_i and p_i denote the position and momentum of the ith particle with respect to an external frame, to the centre-of-mass and relational degrees of freedom $\mathbf{x}_{CMR} := (x_{cm}, p_{cm}, x_{2|1}, p_{2|1})$, where x_{cm}, p_{cm} are the position and momentum of the centre-of-mass with respect to an external frame and $x_{2|1}, p_{2|1}$ are the position and momentum of particle 2 with respect to particle 1, is given by Eq. (6.5) with $N = 2$ and vectors of operators replaced by a single operator. Under this transformation the external covariance matrix $\mathbf{V}_E$ transforms

to $\mathbf{V}_{CMR} = \mathbf{M}_2 \mathbf{V}_E \mathbf{M}_2^T$, where $\mathbf{M}_2$ is given by

$$
\mathbf{M}_2 := \begin{pmatrix} \frac{m_1}{m_1+m_2} & 0 & \frac{m_2}{m_1+m_2} & 0 \\ 0 & 1 & 0 & 1 \\ -1 & 0 & 1 & 0 \\ 0 & -\frac{m_2}{m_1+m_2} & 0 & 1 - \frac{m_2}{m_1+m_2} \end{pmatrix}. \tag{6.24}
$$

As both the external, centre-of-mass, and relational position and momentum operators obey the canonical commutation relations, it follows that $\mathbf{M}_2$ is a symplectic transformation, i.e. it preserves the symplectic form $\mathbf{M}_2 \mathbf{\Omega} \mathbf{M}_2^T = \mathbf{\Omega}$. Since $\mathbf{M}_2$ is symplectic, the associated transformation preserves the Gaussianity of the state, that is, if a state is Gaussian in the external partition, it will also be Gaussian in the centre-of-mass and relational partition.

The relational state ρ_R given in Eq. (6.13) is a Gaussian state whose covariance matrix $\mathbf{V}_{2|1}$ is obtained by deleting the first and second rows and columns of $\mathbf{V}_{CMR}$. Taking the most general form of $\mathbf{V}_1$ and $\mathbf{V}_2$ yields

$$
\mathbf{V}_{2|1} = \frac{1}{\mu_1 \mu_2} \begin{pmatrix} \mu_2 f_1^- + \mu_1 f_2^- & -\mu_2 \tilde{m}_2 g_1 + \mu_1 \tilde{m}_1 g_2 \\ -\mu_2 \tilde{m}_2 g_1 + \mu_1 \tilde{m}_1 g_2 & \mu_2 \tilde{m}_2^2 f_1^+ + \mu_1 \tilde{m}_1^2 f_2^+ \end{pmatrix}, \tag{6.25}
$$

where

$$
f_i^\pm := \cosh 2r_i \pm \cos 2\theta_i \sinh 2r_i,
$$

$$
g_i := \sin 2\theta_i \sinh 2r_i,
$$

and $\tilde{m}_i := m_i/(m_1 + m_2)$.

Entanglement Between the Centre-of-Mass and Relational Degrees of Freedom
As a measure of entanglement we will employ the logarithmic negativity [34]

$$
E_{\mathcal{N}}(\rho) := \log \left\| \rho^{\Gamma_A} \right\|_1, \tag{6.26}
$$

where Γ_A is the partial transpose and $\|\cdot\|_1$ denotes the trace norm, with $\log(\cdot)$ denoting the natural logarithm. The logarithmic negativity is a measure of the failure of the partial transpose of a quantum state to be a valid quantum state and is a faithful measure of entanglement for $1 \times N$ mode Gaussian states [2].

For Gaussian states the logarithmic negativity is given by

$$
E_{\mathcal{N}} := -\sum_k \log \tilde{v}_k \quad \forall \tilde{v}_k < 1, \tag{6.27}
$$

where $\{\tilde{v}_k\}$ is the symplectic spectrum of the partially transposed covariance matrix $\tilde{\mathbf{V}}$, i.e. the eigenspectrum of $|i\mathbf{\Omega}\tilde{\mathbf{V}}|$. The partial transpose of a covariance matrix is

$$
\tilde{\mathbf{V}} = \boldsymbol{\theta}_{1|2} \mathbf{V} \boldsymbol{\theta}_{1|2}, \tag{6.28}
$$

where $\boldsymbol{\theta}_{1|2} = \mathrm{diag}(1, 1, 1, -1)$.

We will use the logarithmic negativity to quantify the entanglement between the centre-of-mass and relational degrees of freedom in $\mathbf{V}_{CMR} = \mathbf{M}_2 \mathbf{V}_E \mathbf{M}_2^T$ for $\mathbf{V}_E = \mathbf{V}_1 \oplus \mathbf{V}_2$, which corresponds to the two particles being in a factorized state $\rho_1 \otimes \rho_2$ in the external partition. $\mathbf{V}_1$ and $\mathbf{V}_2$ will necessarily be of the form given in Eq. (6.23).

Plots of the logarithmic negativity of the state associated with $\mathbf{V}_{CMR}$ for different choices of $\mathbf{V}_1$ and $\mathbf{V}_2$ are given in Figs. 6.1 (identical state parameters), 6.2 (differing purity), and 6.3 (differing squeezing). In Figs. 6.1, 6.2, and 6.3 the dashed red line indicates where the masses of both particles are equal. Several trends emerge from examining these figures.

We first note that equal-mass systems suppress entanglement between the centre-of-mass and relational degrees of freedom. When particles in the external partition are prepared such that they have identical covariance matrices we find vanishing entanglement in the equal-mass case, regardless of the amount of squeezing and rotation. This occurs for both pure and mixed states as illustrated in Figs. 6.1 and 6.2, respectively. As one of the masses gets larger, centre-of-mass and relational entanglement increases for any fixed value of the squeezing parameter r.

Decreasing the purity of the states of the particles in the external partition, shown in Fig. 6.2, indicates the same trends as for the pure case shown in Fig. 6.1. The main effects of decreased purity are to decrease the overall entanglement between the centre-of-mass and relational degrees of freedom and to widen the range of mass ratios for which this entanglement vanishes.

From Figs. 6.1, 6.2, and 6.3 we can observe the effect of phase rotation, corresponding to squeezing along a rotated axis in phase space. For a phase rotation corresponding to $\theta = \theta_1 = \theta_2 = 0$, we find that entanglement between the centre-of-mass and relational degrees of freedom is insensitive to the amount of squeezing. As θ increases we see that squeezing affects this entanglement, particularly as the ratio of the masses increasingly departs from unity. Not surprisingly, entanglement is greater for the pure case, shown in Fig. 6.1, than for the mixed case, shown in Fig. 6.2.

Asymmetric squeezing, $r_2 = \alpha r_1$ where $\alpha \in \mathbb{R}_+$, illustrated in Fig. 6.3 modifies this situation. When there is no squeezing, $r_1 = r_2 = 0$, entanglement between the centre-of-mass and relational degrees of freedom vanishes when the masses of the two particles are equal. However, as r_1 departs from zero the ratio of masses, $m_1/(m_1 + m_2)$, at which this entanglement vanishes increases as illustrated in Fig. 6.3a. This trend is less pronounced as α approaches unity, which is illustrated in Fig. 6.3b. Again, we see that phase rotation plays a significant role; Fig. 6.3c, d demonstrate that if the squeezing of the particles is different and along a rotated axis, entanglement between the centre-of-mass and relational degrees of freedom may not vanish for any mass ratio $m_1/(m_1 + m_2)$.

In Figs. 6.1, 6.2, and 6.3 we have plotted the logarithmic negativity as a measure of the entanglement between the centre-of-mass and relational degrees of freedom for a wide variety of separable states in the external partition. The more entangled these degrees of freedom are, the more mixed the reduced state ρ_R of the relational degrees of freedom will be. The practical consequence of this is that if one wishes to

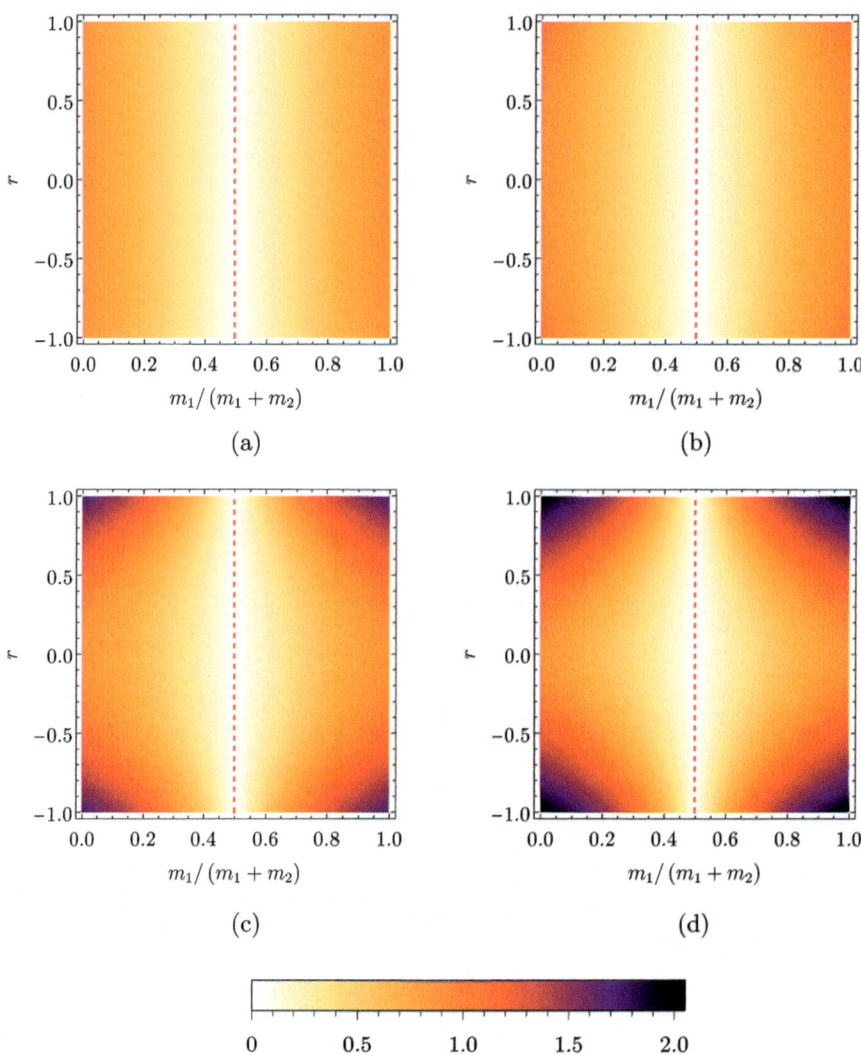

Fig. 6.1 The logarithmic negativity, as a measure of the entanglement between the centre-of-mass and relational degrees of freedom, of the state associated with $\mathbf{V}_{CMR}$, when $\mathbf{V}_1 = \mathbf{V}_2$ and both ρ_1 and ρ_2 are pure, i.e. $\det \mathbf{V}_1 = \det \mathbf{V}_2 = 1$, is plotted as a function of the squeezing parameter $r = r_1 = r_2$ and the ratio of masses $m_1/(m_1 + m_2)$ for different phase rotations $\theta = \theta_1 = \theta_2$: (**a**) $\theta = 0$, (**b**) $\theta = \pi/32$, (**c**) $\theta = \pi/8$, and (**d**) $\theta = \pi/4$

encode quantum information in the relational degrees of freedom of two Gaussian states, perhaps to communicate this information to another party who does not have access to their external reference frame, then the purity and amount and direction of squeezing should be chosen in accordance with Figs. 6.1, 6.2, and 6.3 as to minimize the entanglement between the centre-of-mass and relational degrees of freedom.

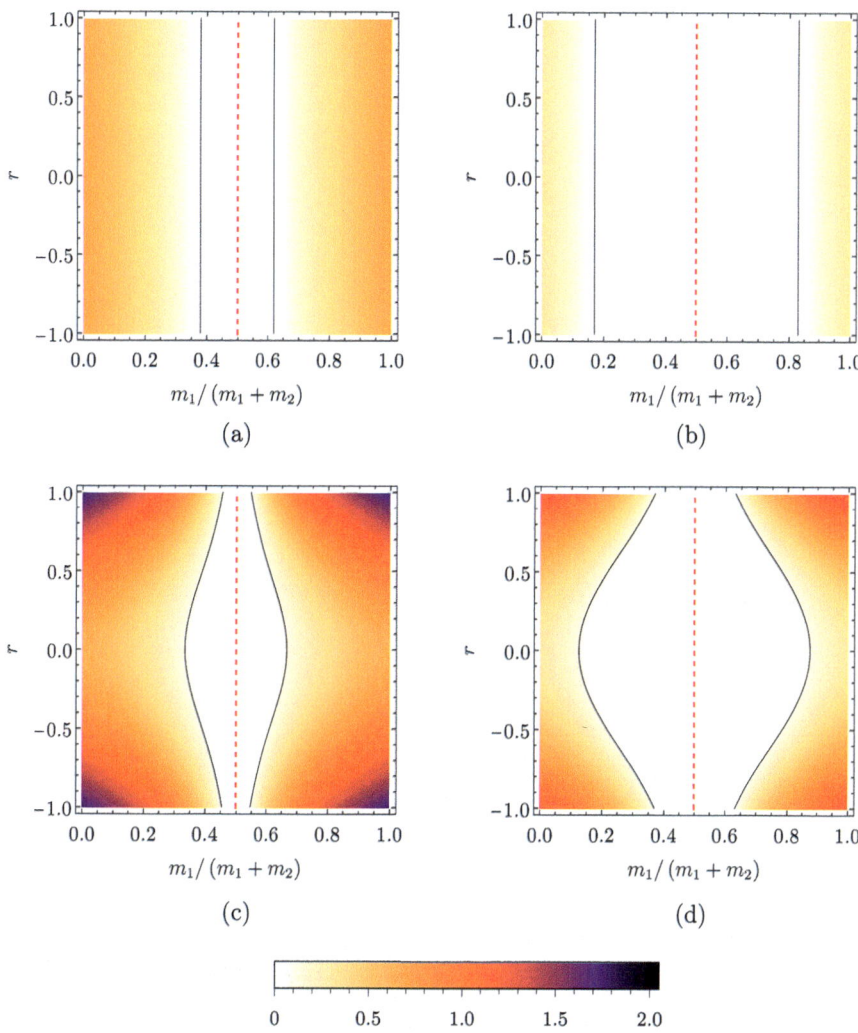

Fig. 6.2 The logarithmic negativity is plotted as a measure of the entanglement between the centre-of-mass and relational degrees of freedom of the state associated with $\mathbf{V}_{CMR}$, with $r = r_1 = r_2$ and $\theta = \theta_1 = \theta_2$, for different purities μ_1 of particle 1 and phase rotations θ; the state of particle 2 is pure, $\mu_2 = 1$. In (**a, b**) $\theta = 0$ and (**c, d**) $\theta = \pi/4$. In (**a, c**) $\mu_1 = 0.6$ and (**b, d**) $\mu = 0.2$. Plots for $\theta = 0$ and $\mu_1 = 1$ and for $\theta = \pi/4$ and $\mu_1 = 1$ are shown in Fig. 6.1a and d, respectively

6.2.3 Purity of the Relational State

As considered above, particles 1 and 2 are prepared in a pure factorized state in the external partition, and since the transformation to the centre-of-mass and relational

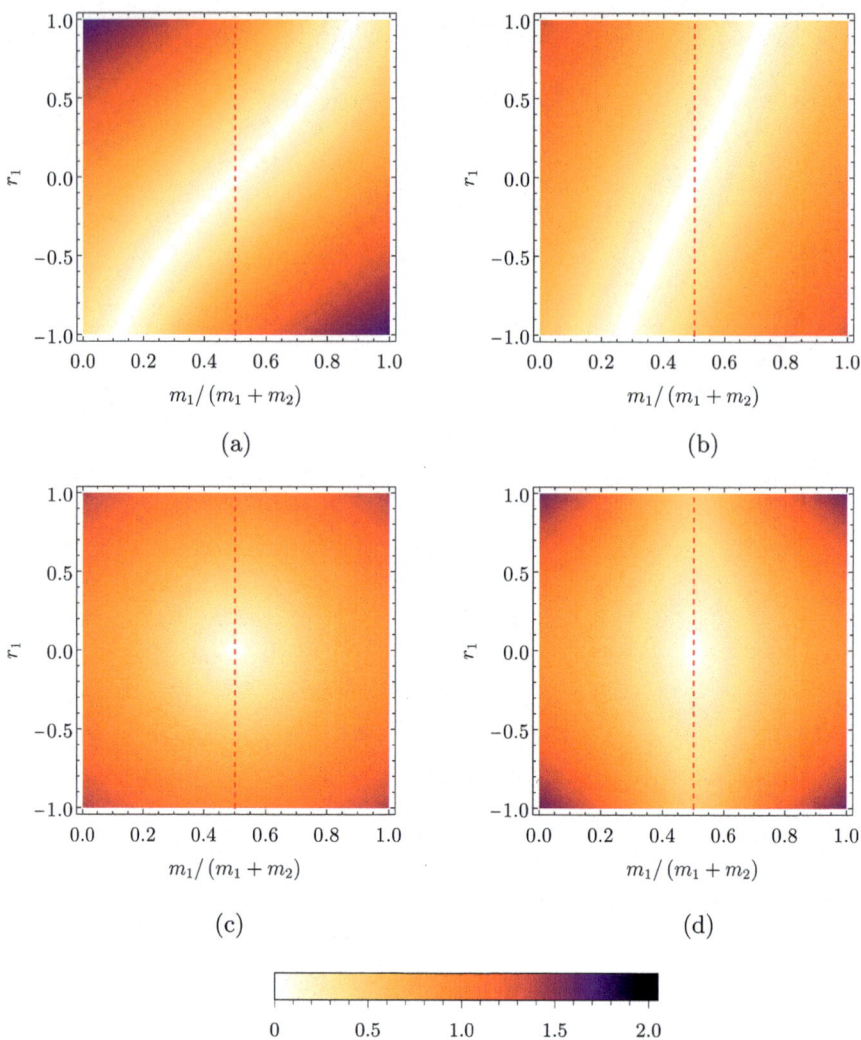

Fig. 6.3 The logarithmic negativity of the state associated with $\mathbf{V}_{CMR}$ is plotted as a measure of the entanglement between the centre-of-mass and relational degrees of freedom, with $\det \mathbf{V}_1 = \det \mathbf{V}_2 = 1$ and $r_2 = \alpha r_1$, for different phase rotations $\theta = \theta_1 = \theta_2$ and values of α. In (**a, b**) $\theta = 0$ and (**c, d**) $\theta = \pi/4$. In (**a, c**) $\alpha = 0$ and (**b, d**) $\alpha = 0.5$. Plots for $\theta = 0$ and $\alpha = 1$ and for $\theta = \pi/4$ and $\alpha = 1$ are shown in Fig. 6.1a and d, respectively

partition is a unitary operation, the purity of the reduced state on the relational degrees of freedom ρ_R quantifies how much information about the external frame has been lost in tracing out the centre-of-mass degrees of freedom. We explicitly compute the purity of the relational state ρ_R given the joint state in the external partition is pure and factorized and both particles 1 and 2 are in Gaussian states.

The covariance matrices considered in Sect. 6.2.2 were of the form $\mathbf{V}_E = \mathbf{V}_1 \oplus \mathbf{V}_2$, where both $\mathbf{V}_1$ and $\mathbf{V}_2$ were given by Eq. (6.23). The purity of $\mathbf{V}_{CMR} = \mathbf{M}_2 \mathbf{V}_E \mathbf{M}_2^T$ is given by

$$\mu_{CMR} = \frac{1}{\sqrt{\det \mathbf{V}_{CMR}}} = \mu_1 \mu_2, \tag{6.29}$$

where μ_1 and μ_2 are the purities associated with $\mathbf{V}_1$ and $\mathbf{V}_2$, respectively.

The purity of the relational state defined by $\mathbf{V}_{2|1}$ in Eq. (6.25), that is, the state obtained from $\mathbf{V}_{CMR}$ by taking the partial trace over the centre-of-mass degrees of freedom, is

$$
\begin{aligned}
\mu_{2|1} &= \frac{1}{\sqrt{\det \mathbf{V}_{2|1}}} \\
&= \mu_1 \mu_2 \Big[\mu_2^2 \tilde{m}_2^2 f_1^- f_1^+ + \mu_1 \mu_2 \left(\tilde{m}_1^2 f_1^- f_2^+ + \tilde{m}_2^2 f_1^+ f_2^- \right) + \mu_1^2 \tilde{m}_1^2 f_2^- f_2^+ \\
&\quad - \mu_2^2 \tilde{m}_2^2 g_1^2 + 2\mu_1 \mu_2 \tilde{m}_1 \tilde{m}_2 g_1 g_2 - \mu_1^2 \tilde{m}_1^2 g_2^2 \Big]^{-1/2}, \tag{6.30}
\end{aligned}
$$

where we have introduced the notation $\tilde{m}_i := m_i/(m_1 + m_2)$.

If $\mathbf{V}_{CMR}$ is pure, which corresponds to both $\mathbf{V}_1$ and $\mathbf{V}_2$ being pure, then $\mu_{CMR} = 1$ and $\mu_{2|1}$ is a genuine measure of entanglement between the centre-of-mass and relational degrees of freedom. In this case, $\mu_{2|1}^{-2}$ simplifies to

$$
\begin{aligned}
\mu_{2|1}^{-2} &= (\tilde{m}_2 - \tilde{m}_1) \Big[\sinh(2r_1) \cosh(2r_2) \cos(2\theta_1) - \sinh(2r_2) \cosh(2r_1) \cos(2\theta_2) \Big] \\
&\quad - \sinh(2r_1) \sinh(2r_2) \Big[2\tilde{m}_1 \tilde{m}_2 \cos[2(\theta_1 + \theta_2)] + \cos(2\theta_1) \cos(2\theta_2) \Big] \\
&\quad + (2\tilde{m}_1 \tilde{m}_2 + 1) \cosh(2r_1) \cosh(2r_2) + \tilde{m}_1^2 + \tilde{m}_2^2. \tag{6.31}
\end{aligned}
$$

If the mass of the two particles are equal $m_1 = m_2$, $\mu_{2|1}^{-2}$ further simplifies to

$$
\begin{aligned}
\mu_{2|1}^{-2} &= \frac{1}{4} \Big[-2 \sinh(2r_1) \sinh(2r_2) \cos[2(\theta_1 - \theta_2)] \\
&\quad + \cosh[2(r_1 - r_2)] + \cosh[2(r_1 + r_2)] + 2 \Big]. \tag{6.32}
\end{aligned}
$$

From Eq. (6.32), we observe that when the masses of the two particles are identical $m_1 = m_2$, and each particle is squeezed by the same amount $r_1 = r_2$ and in the same direction in phase space $\theta_1 = \theta_2$, the reduced state associated with the covariance matrix $\mathbf{V}_{2|1}$ is pure, i.e. $\mu_{2|1} = 1$, which corresponds to vanishing entanglement between the centre-of-mass and relational degrees of freedom. This agrees with the plots of the logarithmic negativity in Fig. 6.1.

For the case when $m_1 \neq m_2$, $r_1 = r_2 = r$ and $\theta_1 = \theta_2 = \theta$, corresponding to Fig. 6.1, $\mu_{2|1}^{-2}$ becomes

$$\mu_{2|1}^{-2} = 2\frac{m_1^2 + m_2^2}{(m_1 + m_2)^2} + \sin^2(2\theta)\left[\frac{m_1^2 + m_2^2}{(m_1 + m_2)^2}\sinh^2(2r) - 2\frac{m_1 m_2}{(m_1 + m_2)^2}\right].$$

(6.33)

When the mass of either particle becomes infinite we find

$$\mu_{2|1}^{-2} = 2 + \sinh^2(2r)\cos^2(2\theta).$$

(6.34)

6.2.4 The Three-Particle Case

We now consider a similar analysis for a system of three particles with masses m_1, m_2, and m_3. When transforming a fully factorized state in the external partition $\mathcal{H} = \mathcal{H}_1 \otimes \mathcal{H}_2 \otimes \mathcal{H}_3$ to the centre-of-mass and relational partition $\mathcal{H} = \mathcal{H}_{CM} \otimes \mathcal{H}_R$, there will again be entanglement generated between the centre-of-mass and relational degrees of freedom. In addition, there will be entanglement generated among the relational degrees of freedom, a new feature not possible for the two-particle system considered above.

The centre-of-mass position and momentum operators, along with the relative position and momentum operators, are again defined via Eq. (6.5). The transformed covariance matrix is given by $\mathbf{V}_{CMR} = \mathbf{M}_3 \mathbf{V}_E \mathbf{M}_3^T$, where

$$\mathbf{M}_3 := \begin{pmatrix} \frac{m_1}{M} & 0 & \frac{m_2}{M} & 0 & \frac{m_3}{M} & 0 \\ 0 & 1 & 0 & 1 & 0 & 1 \\ -1 & 0 & 1 & 0 & 0 & 0 \\ 0 & -\frac{m_2}{M} & 0 & 1-\frac{m_2}{M} & 0 & -\frac{m_2}{M} \\ -1 & 0 & 0 & 0 & 1 & 0 \\ 0 & -\frac{m_3}{M} & 0 & -\frac{m_3}{M} & 0 & 1-\frac{m_3}{M} \end{pmatrix}.$$

(6.35)

The relational state defined by $\mathbf{V}_{23|1}$ of particles 2 and 3 as described by particle 1 is obtained by deleting the first and second rows and columns of $\mathbf{V}_{CMR}$. We observe that in the limit when m_3 vanishes and the columns and rows of $\mathbf{M}_3$ associated with particle 3 are deleted (the last two rows and columns), $\mathbf{M}_2$ as defined in Eq. (6.24) is recovered.

We assume the state of the three-particle system in the external partition is a fully factorized Gaussian state with the covariance matrix $\mathbf{V}_E = \mathbf{V}_1 \oplus \mathbf{V}_2 \oplus \mathbf{V}_3$. For simplicity we restrict ourselves to the case when $\mathbf{V}_1 = \mathbf{V}_2 = \mathbf{V}_3$ and $\det \mathbf{V}_E = 1$, in other words, a pure state with each of the three particles identically squeezed in the same direction.

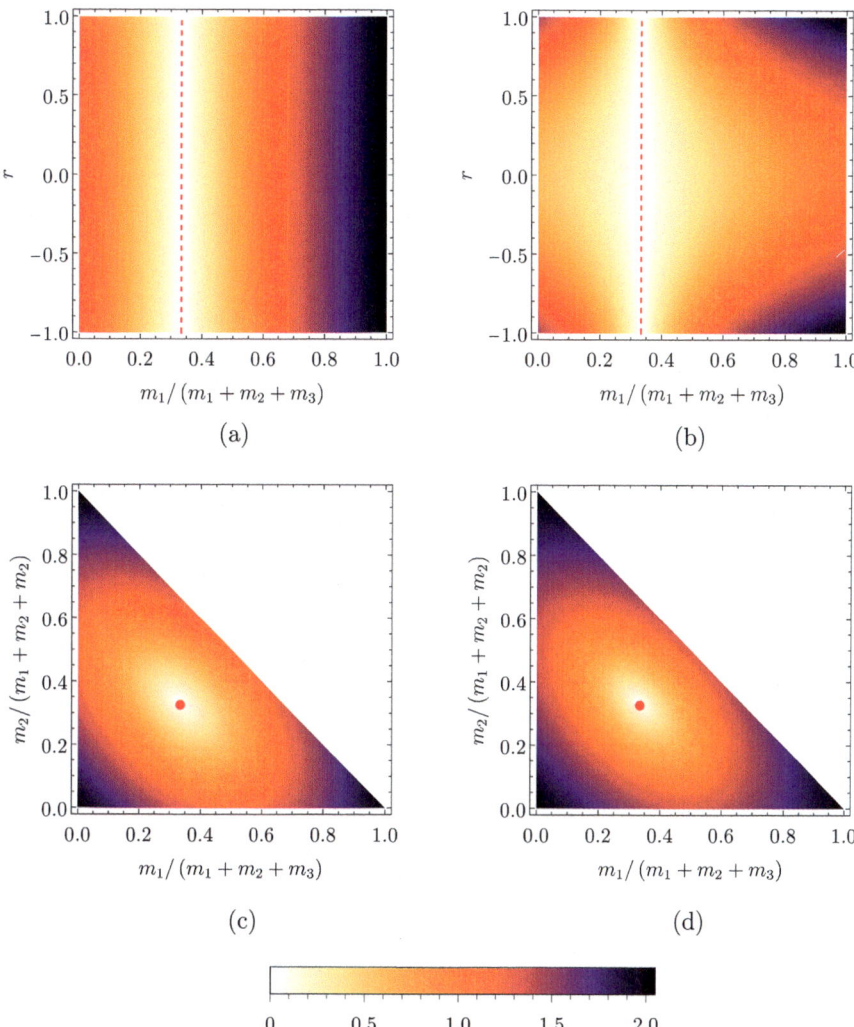

Fig. 6.4 As a measure of the entanglement between the centre-of-mass and relational degrees of freedom in the three-particle case, the logarithmic negativity of the state associated with $\mathbf{V}_{CMR}$ is plotted for different equal phase rotations $\theta_1 = \theta_2 = \theta_3 = \theta$ with $\det \mathbf{V}_1 = \det \mathbf{V}_2 = \det \mathbf{V}_3 = 1$. In (**a, b**) the logarithmic negativity is plotted for the case when $m_2 = m_3$ as a function of $m_1/(m_1 + m_2 + m_3)$ and equal squeezing parameter $r_1 = r_2 = r_3 = r$, with $\theta = 0$ and $\theta = \pi/4$, respectively. In (**c, d**) the logarithmic negativity is plotted as a function of the two mass ratios $m_1/(m_1 + m_2 + m_3)$ and $m_2/(m_1 + m_2 + m_3)$ for $\theta = 0$ and $\theta = \pi/4$, respectively, with the equal squeezing parameter fixed at $r = 0.7$

In Fig. 6.4 the logarithmic negativity as a measure of entanglement between the centre-of-mass and relational degrees of freedom in $\mathbf{V}_{CMR}$ is plotted for different choices of $\mathbf{V}_E$. In Fig. 6.5 the logarithmic negativity between the relational degrees

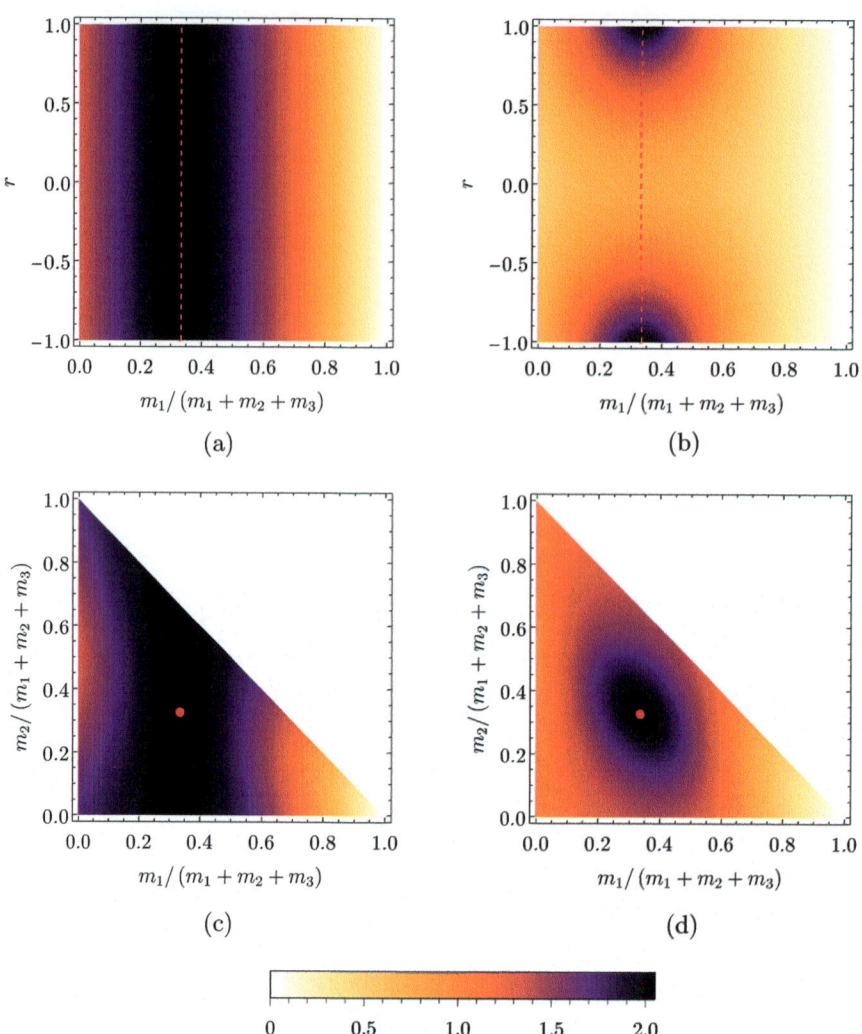

Fig. 6.5 The logarithmic negativity of the relative state of particles 2 and 3 defined by $\mathbf{V}_{23|1}$ is plotted, characterizing the entanglement among the relational degrees of freedom, for different equal phase rotations $\theta_1 = \theta_2 = \theta_3 = \theta$ with $\det \mathbf{V}_1 = \det \mathbf{V}_2 = \det \mathbf{V}_3 = 1$. In (**a, b**) the logarithmic negativity is plotted for the case $m_2 = m_3$ as a function of the ratio $m_1/(m_1 + m_2 + m_3)$ and equal squeezing parameter $r_1 = r_2 = r_3 = r$ for $\theta = 0$ and $\theta = \pi/4$, respectively. In (**c, d**) logarithmic negativity is plotted as a function of the mass ratios $m_1/(m_1 + m_2 + m_3)$ and $m_2/(m_1 + m_2 + m_3)$ for equal squeezing parameter $r = 0.7$ and $\theta = 0$ and $\theta = \pi/4$, respectively

of freedom in $\mathbf{V}_{23|1}$ is plotted for different choices of $\mathbf{V}_E$. In both figures, the dashed red line and red point indicate where the masses of all particles are equal.

In the three-particle case, entanglement between the centre-of-mass and relational degrees of freedom behaves similar to the two-particle case. However, entanglement among the relational degrees of freedom—in the case at hand, the entanglement between particles 2 and 3 as described by particle 1—exhibits strikingly different behaviour; this is illustrated in Fig. 6.5. Such entanglement is maximized in the equal-mass case, shown in Fig. 6.5b, d, provided there is some phase rotation. In the absence of phase rotation, this effect vanishes. For all values of the (equal) phase rotation parameter, we observe that when the mass of the reference particle m_1 becomes infinite, the entanglement between particles 2 and 3 vanishes. This is as expected, since this limit corresponds to particle 1 behaving as a classical reference frame with a large mass. Indeed, we noted that in the limit $m_1 \rightarrow \infty$, the 4×4 lower-right submatrix of $\mathbf{M}_3$ becomes the identity matrix, and the only effect of the change of coordinates is that of redefining the origin in space for the coordinates of the second and third particle.

6.3 Summary

In this chapter we have highlighted issues involving quantum reference frames associated with noncompact groups. We began in Sect. 6.1.1 by introducing the usually employed G-twirl as a relational description between quantum systems and demonstrated how it leads to non-normalizable states when averaging states over noncompact groups. In Sect. 6.1.2 we demonstrated that a lack of reference frame associated with the translation group and the group of Galilean boosts leads to a superselection rule on the respective momentum and position of the centre-of-mass of a multiparticle system. Further, we saw how the G-twirl over these groups leads to the appearance of the reduced state on the relational degrees of freedom previously considered by Angelo et al. [11]. We then examined the consequences of this relational description in Sect. 6.2 by studying the entanglement that emerges between the centre-of-mass degrees of freedom and the relational degrees of freedom, as well as the entanglement among the relational degrees of freedom, for a system of particles when moving from a description of the quantum system entirely with respect to an external frame, to a description in which only the centre-of-mass is specified with respect to an external frame and all other degrees of freedom are relational.

Two main observations emerged from studying the reduced state ρ_R on the relational degrees of freedom, introduced in Eq. (6.13), for systems of two and three particles. First, for fully separable Gaussian states in the external partition with identical second moments, entanglement between the centre-of-mass degrees of freedom and relational degrees of freedom is minimized when the masses of the particles are the same. Second, again for fully separable Gaussian states in the external partition with identical second moments, in the limit when the mass of the reference particle (the particle for which the relational degrees of freedom are defined with respect to) becomes infinite, the entanglement among the relational degrees of freedom vanishes. This second observation suggests a meaningful way to interpret the external reference frame, with respect to which we usually describe a quantum state, as the limit of a physical system, say a particle, in which its mass is taken to infinity [6]. The consequences of this second observation will be explored in future work.

It may be possible to gain further physical intuition into the behaviour of the informational properties of ρ_R by comparing ρ_R with the behaviour of nonclassical states of light passing through a beam splitter, as this scenario has been well studied in the field of quantum optics [21, 32] and the formalism of Gaussian quantum information was developed with this situation in mind [1, 35].

The primary motivation for examining quantum reference frames associated with noncompact groups is to apply the quantum reference frame formalism to relativistic systems, in which the group associated with changes of a reference frame is the Poincaré group. We note that the approach taken in Sect. 6.1.2 was to introduce the relative and centre-of-mass partition of the Hilbert space and then show that the relative degrees of freedom form a decoherence-free subsystem, whereas the centre-of-mass degree of freedom forms a decoherence-full subsystem; see Eq. (6.12). This approach may not be possible for the Poincaré group as the usually defined centre-of-mass is not covariant [4]. In this case, the decoherence-free and decoherence-full subspaces will need to be identified from the structure of the Poincaré group [18].

It will also be interesting to explore whether it is possible to construct a relational quantum theory, similar to what was done in Ref. [27], for the Galilean group using the relational description in Eq. (6.13) and examine how the usual "nonrelational" theory emerges.

References

1. G. Adesso, F. Illuminati, Entanglement in continuous variable systems: recent advances and current perspectives. J. Phys. A **40**, 7821 (2007)
2. G. Adesso, A. Serafini, F. Illuminati, Quantification and scaling of multipartite entanglement in continuous variable systems. Phys. Rev. Lett. **93**, 220504 (2004)
3. G. Adesso, S. Ragy, A.R. Lee, Continuous variable quantum information: Gaussian states and beyond. Open Syst. Inf. Dyn. **21**, 1440001 (2014)
4. P. Aguilar, C. Chryssomalakos, H.H. Coronado, E. Okon, Position operators and center of mass: new perspectives. Int. J. Mod. Phys. A **28**, 1350146 (2013)
5. Y. Aharonov, G. Carmi, Quantum aspects of the equivalence principle. Found. Phys. **3**, 493 (1973)
6. Y. Aharonov, T. Kaufherr, Quantum frames of reference. Phys. Rev. D **30**, 368 (1984)
7. Y. Aharonov, L. Susskind, Charge superselection rule. Phys. Rev. **155**, 1428 (1967)
8. Y. Aharonov, L. Susskind, Observability of the sign change of spinors. Phys. Rev. **158**, 1428 (1967)
9. M. Ahmadi, A.R.H. Smith, A. Dragan, Communication between inertial observers with partially correlated reference frames. Phys. Rev. A **92**, 062319 (2015)
10. R.M. Angelo, A.D. Ribeiro, Kinematics and dynamics in noninertial quantum frames of reference. J. Phys. A **45**, 465306 (2012)

11. R.M. Angelo, N. Brunner, S. Popescu, A.J. Short, P. Skrzypczyk, Physics within a quantum reference frame. J. Phys. A **44**, 145304 (2011)
12. S.D. Bartlett, T. Rudolph, R.W. Spekkens, Reference frames, superselection rules, and quantum information. Rev. Mod. Phys. **79**, 555 (2007)
13. S.D. Bartlett, T. Rudolph, R.W. Spekkens, P.S. Turner, Quantum communication using a bounded-size quantum reference frame. New J. Phys. **11**, 063013 (2009)
14. A. Chęcińska, A. Dragan, Communication between general-relativistic observers without a shared reference frame. Phys. Rev. A **92**, 012321 (2015)
15. M.R. Dowling, S.D. Bartlett, T. Rudolph, R.W. Spekkens, Observing a coherent superposition of an atom and a molecule. Phys. Rev. A **74**, 052113 (2006)
16. D. Giulini, States, symmetries and superselection, in *Decoherence: Theoretical, Experimental, and Conceptual Problems*, ed. by P. Blanchard, E. Joos, D. Giulini, C. Kiefer, I.-O. Stamatescu (Springer, Berlin, 2000), pp. 87–100
17. M. Jarzyna, R. Demkowicz-Dobrzański, Quantum interferometry with and without an external phase reference. Phys. Rev. A **85**, 011801 (2012)
18. O. Kabernik, *Quantum Reference Frames and the Poincaré Symmetry*, Master's thesis, University of Waterloo, 2014
19. A. Kitaev, D. Mayers, J. Preskill, Superselection rules and quantum protocols. Phys. Rev. A **69**, 052326 (2004)
20. E. Lubkin, On violation of the superselection rules. Ann. Phys. **56**, 69 (1970)
21. L. Mandel, E. Wolf, *Optical Coherence and Quantum Optics* (Cambridge University Press, Cambridge, 1995)
22. I. Marvian, R.B. Mann, Building all time evolutions with rotationally invariant Hamiltonians. Phys. Rev. A **78**, 022304 (2008)
23. I. Marvian, R.W. Spekkens, The theory of manipulations of pure state asymmetry: I. Basic tools, equivalence classes and single copy transformations. New J. Phys. **15**, 033001 (2013)
24. R. Mirman, Coherent superposition of charge states. Phys. Rev. **186**, 1380 (1969)
25. R. Mirman, Analysis of the experimental meaning of coherent superposition and the nonexistence of superselection rules. Phys. Rev. D **1**, 3349 (1970)
26. M.C. Palmer, F. Girelli, S.D. Bartlett, Changing quantum reference frames. Phys. Rev. A **89**, 052121 (2014)
27. D. Poulin, Toy model for a relational formulation of quantum theory. Int. J. Theor. Phys. **45**, 1189 (2006)
28. C. Rovelli, Quantum reference systems. Classical Quantum Gravity **8**, 317 (1991)
29. C. Rovelli, What is observable in classical and quantum gravity? Classical Quantum Gravity **8**, 297 (1991)
30. C. Rovelli, Relational quantum mechanics. Int. J. Theor. Phys. **35**, 1637 (1996)
31. C. Rovelli, *Quantum Gravity* (Cambridge University Press, Cambridge, 2004)
32. M.O. Scully, M.S. Zubairy, *Quantum Optics* (Cambridge University Press, Cambridge, 1997)
33. L. Smolin, The structural foundations of quantum gravity, in *The Case for Background Independence* (Oxford University Press, Oxford, 2006), pp. 196–239
34. G. Vidal, R.F. Werner, A computable measure of entanglement. Phys. Rev. A **65**, 032314 (2002)
35. C. Weedbrook, S. Pirandola, R. García-Patrón, N.J. Cerf, T.C. Ralph, J.H. Shapiro, S. Lloyd, Gaussian quantum information. Rev. Mod. Phys. **84**, 621 (2012)
36. J. Williamson, On the algebraic problem concerning the normal forms of linear dynamical systems. Am. J. Math. **58**, 141 (1936)

Chapter 7
Communication Without a Shared Reference Frame

Most quantum communication protocols assume that the parties communicating share a classical background reference frame. For example, suppose Alice wishes to communicate to Bob the state of a qubit using a teleportation protocol [9]. Alice begins by having the qubit she wishes to communicate to Bob interact with one half of an entangled pair of qubits shared by her and Bob. Alice then measures the two qubits in her possession and picks up the phone and informs Bob of the measurement result. Bob uses this information to apply an appropriate gate to his half of the entangled pair to recover the state Alice wished to send to him.

The success of this protocol depends on Alice's ability to classically communicate to Bob which gates he should apply to his half of the entangled state. This can only be done if Alice and Bob share a reference frame. As an example, suppose Alice informs Bob that he needs to apply the Pauli z operator to the qubit in his possession. If Bob is ignorant of the orientation of his laboratory with respect to Alice's, he does not know in which direction to orientate the magnetic field in his Stern-Gerlach apparatus to implement the Pauli z operator to recover the state sent by Alice. In this case the teleportation protocol is unable to be carried out perfectly [5, 15, 16].

This motivates the study of quantum communication without a shared reference frame [2]. One way Alice can communicate to Bob, despite not sharing a reference frame with him, is to encode information into degrees of freedom that are invariant under a change of Alice's reference frame. Without knowing his relation to Alice's reference frame, Bob is able to extract both classical and quantum information encoded in these degrees of freedom [1]. However, in practice such communication schemes may be challenging to implement since they require highly entangled states of many qubits.

Another possibility for Alice and Bob to communicate without a shared reference frame is for Alice to send Bob a quantum system ρ_R to serve as a token of her reference frame, together with the state ρ_S she wishes to communicate to him. Since Bob does not know the relation between his reference frame and Alice's,

© Springer Nature Switzerland AG 2019   121
A. R. H. Smith, *Detectors, Reference Frames, and Time*, Springer Theses,
https://doi.org/10.1007/978-3-030-11000-0_7

with respect to his reference frame he will see the joint state $\rho_R \otimes \rho_S$ averaged over all possible orientations of his laboratory with respect to Alice's; this averaging operation is referred to as the G-twirl and the averaged state denoted by $\mathcal{G}[\rho_R \otimes \rho_S]$. Bob can apply a recovery operation to this G-twirled state by measuring the reference token and applying an appropriate correction to the system Alice wishes to send to him, allowing him to recover a state ρ_S' that is close to ρ_S. This recovery operation was first constructed by Bartlett et al. [3], and its success was found to depend on the size of the reference token, which is necessarily bounded if the reference token is described by a finite dimensional Hilbert space.

However, this communication protocol is based on Bob assigning the G-twirled state $\mathcal{G}[\rho_R \otimes \rho_S]$ to the system and reference token, and the G-twirl does not yield normalizable states when the group of reference frames being averaged over in noncompact [12]. This begs the question: Can an analogous communication protocol involving a reference token sent by Alice and a recovery operation implemented by Bob be constructed given that changes of their reference frames form a noncompact group? Furthermore, if the Hilbert space of the reference token is infinite dimensional, for example, $\mathcal{H}_R \simeq L^2(\mathbb{R})$, what physical aspect of the reference token acts as its size?

The purpose of this chapter is to examine these questions. Considerations of noncompact groups within the theory of quantum reference frames is important if one hopes to apply the theory to the physically relevant Galilean and Poincaré groups, which are both noncompact.

We begin in Sect. 7.1 by describing the encoding and recovery operations introduced by Bartlett et al. [3]. In Sect. 7.2 we introduce a G-twirl over a compact subset of a noncompact group and a complementary recovery operation such that in the limit when this G-twirl becomes an average over the entire noncompact group, the composition of the recovery operation with this G-twirl results in properly normalized states. We then apply this construction in Sect. 7.3 to the case when Alice and Bob do not share a reference frame associated with the one-dimensional translation group, which is relevant for parties communicating without a shared positional reference frame. In this case, we identify the inverse of the width in position space of the reference token's state as the size of the reference token, and demonstrate that in the limit when this width goes to zero Alice and Bob are able to communicate perfectly without a shared reference frame. We conclude in Sect. 7.4 with a summary of the results presented in this chapter.

7.1 Communication Without a Shared Classical Reference Frame

Consider two parties, Alice and Bob, each employing their own classical reference frame to describe the state of a single quantum system associated with the Hilbert space $\mathcal{H}_S$. Suppose that this system transforms via a unitary representation of the group G when changing the reference frame used to describe the system; for the time being we will assume G is a compact Lie group.

Let $g \in G$ label the group element which describes the transformation from Alice's to Bob's reference frame. If Alice prepares the system in the state $\rho_S \in \mathcal{S}(\mathcal{H}_S)$ with respect to her reference frame, where $\mathcal{S}(\mathcal{H}_S)$ is the space of states on $\mathcal{H}_S$ and g is completely unknown to Bob, then the state with respect to his reference frame will be given by a uniform average over all possible $g \in G$; that is, by the G-twirl

$$\mathcal{G}[\rho_S] := \int_G dg \, U_S(g) \, \rho_S \, U_S(g)^\dagger, \tag{7.1}$$

where dg denotes the Haar measure associated with G and $U_S(g) \in \mathcal{U}(\mathcal{H}_S)$ is the unitary representation of the group element $g \in G$ on $\mathcal{H}_S$, with $\mathcal{U}(\mathcal{H}_S)$ denoting the space of unitary operators on $\mathcal{H}_S$. If instead Bob has some partial information about the relation between his reference frame and Alice's, the uniform average over all possible $g \in G$ in Eq. (7.1) would be replaced with a weighted average encoding Bob's partial information.

In general, the G-twirl results in decoherence, not from the system interacting with an environment and information being lost to the environment, but from Bob's lack of knowledge about the relationship between his reference frame and Alice's. To combat this decoherence, Alice may prepare another quantum system, described by the Hilbert space $\mathcal{H}_R$, to serve as a token of her reference frame (a good representative of her reference frame). Suppose Alice prepares the token in the state $|e\rangle \in \mathcal{H}_R$, then the reference token and system relative to Bob's frame will be given by the encoding operation

$$\begin{aligned} \mathcal{E}[\rho_S] :=& \mathcal{G}\big[|e\rangle\langle e| \otimes \rho_S \big] \\ =& \int_G dg \, \mathcal{U}_R(g)[|e\rangle\langle e|] \otimes \mathcal{U}_S(g)[\rho_S], \end{aligned} \tag{7.2}$$

where $\mathcal{U}_i(g)[\rho] := U_i(g) \, \rho \, U_i(g)^\dagger$ denotes the adjoint representation of the action of the group element $g \in G$ on $\rho \in \mathcal{S}(\mathcal{H}_i)$ for $i \in \{R, S\}$.

Bob's task is now to best recover the state of the system ρ_S given the encoded state $\mathcal{E}[\rho_S]$. In other words, he must construct a recovery operation

$$\mathcal{R} : \mathcal{S}(\mathcal{H}_R \otimes \mathcal{H}_S) \to \mathcal{S}(\mathcal{H}_S), \tag{7.3}$$

that when applied to $\mathcal{E}[\rho_S]$ results in a state $\rho_S' \in \mathcal{S}(\mathcal{H}_S)$ that is as close as possible to ρ_S. A recovery operation $\mathcal{R}$ was constructed by Bartlett et al. [3] with such properties, and its action on the encoded state $\mathcal{E}[\rho_S]$ yields

$$\rho_S' := \mathcal{R} \circ \mathcal{E}[\rho_S] = \int_G dg \, p(g) \mathcal{U}_S(g)[\rho_S], \tag{7.4}$$

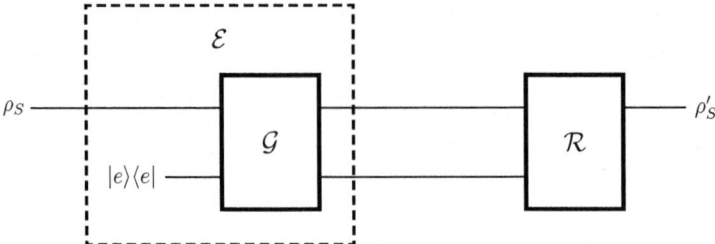

Fig. 7.1 The communication channel $\mathcal{R} \circ \mathcal{E}$. Alice prepares a state ρ_S she wishes to communicate to Bob along with the state $|e\rangle\langle e|$ as a token of her reference frame. As Bob does not know the relation between his reference frame and Alice's, he sees the joint state of the reference token and system as the encoded state $\mathcal{E}[\rho_S] = \mathcal{G}\big[|e\rangle\langle e| \otimes \rho_S\big]$. Bob then applies the recovery operation to the encoded state and recovers the state $\rho_S' = \mathcal{R} \circ \mathcal{E}[\rho_S]$

where $p(g) \propto |\langle e|U_R(g)|e\rangle|^2$ with $U_R(g) \in \mathcal{U}(\mathcal{H}_R)$ being the unitary representation of $g \in G$ on $\mathcal{H}_R$. We will explicitly construct this recovery operation in the next section for the case when G is noncompact (Fig. 7.1).

7.2   A Recovery Operation for Noncompact Groups

The action of the G-twirl over a noncompact group on a state results is a non-normalizable density matrix [12], and therefore it is not clear whether the encoding operation $\mathcal{E}$ or the recovery operation $\mathcal{R}$ discussed above are applicable to reference frames associated with noncompact groups. We now demonstrate that despite this fact, the composition of an encoding operation associated with a noncompact group with a suitably defined recovery operation results in a properly normalized state.

The approach we will take is to define a G-twirl over a compact subset of the noncompact group associated with the reference frame, which corresponds to Bob having partial information that the relation between his reference frame and Alice's is described by $g \in [-\tau, \tau] \subset G$. This finite G-twirl will be used in an encoding operation analogous to Eq. (7.2). We will then construct a complementary recovery operation, compose it with this encoding operation (similar to Eq. (7.4)), and finally take a limit in which the finite G-twirl corresponds to twirling over the entire noncompact group. We will show that in this limit the recovered state is properly normalized.

7.2.1   The Encoding Map

Consider all possible transformations of Alice's and Bob's classical reference frames to form a strongly continuous one-parameter noncompact Lie group G. Suppose that the unitary representation of a group element $g \in G$ on the Hilbert space $\mathcal{H}_R$ describing the reference token is $U_R(g) \in \mathcal{U}(\mathcal{H}_R)$. By Stone's theorem [13], $U_R(g) = e^{igA_R}$ is generated by a self-adjoint operator A_R, the spectrum of which we denote by $\sigma(A_R)$ and assume to be continuous.[1] For each element of the spectrum $f(a_R) \in \sigma(A_R)$ there corresponds an eigenket $|a_R\rangle$ such that

$$A_R |a_R\rangle = f(a_R) |a_R\rangle, \tag{7.5}$$

with eigenvalue $f(a_R) \in \mathbb{R}$. Since $\sigma(A_R)$ is continuous and A_R is self-adjoint, these eigenkets are normalized with the Dirac delta function

$$\langle a_R | a_R' \rangle = \delta(a_R - a_R'). \tag{7.6}$$

From the above normalization condition we see that $|a_R\rangle \notin \mathcal{H}_R$, as these eigenkets are not square integrable and therefore do not represent physical states.[2]

Our first step is to construct a well-defined encoding operation analogous to Eq. (7.2). To do so, we suppose the state of Alice's reference token $|e\rangle \in \mathcal{H}_R$, expressed in the basis furnished by the eigenkets of A_R, is

$$|e\rangle := \int da_R \, \psi_R(a_R) |a_R\rangle, \tag{7.7}$$

where $\psi_R(a_R) := \langle a_R | e \rangle$. Next, let us introduce the set of states

$$\left\{ |e(g)\rangle := U_R(g) |e\rangle \,\, \big| \,\, \forall g \in G \right\}, \tag{7.8}$$

where each $|e(g)\rangle$ corresponds to a different orientation of Alice's reference frame. The state of the reference token $|e\rangle$ should be chosen such that each $|e(g)\rangle$ defined in Eq. (7.8) is distinct, that is, the state of the reference token should not be invariant with respect to G. Furthermore, for the states $|e(g)\rangle$ to imitate a classical reference frame, they must be orthogonal so as they are perfectly distinguishable.

Now suppose Alice prepares her reference token in the state $\rho_R \in \mathcal{S}(\mathcal{H}_R)$ and wishes to send Bob the state $\rho_S \in \mathcal{S}(\mathcal{H}_S)$ of a system associated with the Hilbert space $\mathcal{H}_S$. If Bob knows the relation between his reference frame and Alice's is given by a group element $g \in [-\tau, \tau] \subset G$, but within this interval he is completely

---

[1] This is true of the group generated by either the position or momentum operator on $L^2(\mathbb{R})$. We note that the following construction does not rely on $\sigma(A_R)$ being continuous.

[2] More precisely, when dealing with operators with a continuous spectrum the theory is defined on a rigged Hilbert space [6]

$$\Phi \subset \mathcal{H}_R \subset \Phi',$$

where Φ is a proper subset dense in $\mathcal{H}_R$ and Φ' is the dual of Φ, defined through the inner product on $\mathcal{H}_R$. In our case, Φ is the Schwarz space of smooth rapidly decreasing functions on $\mathbb{R}$ and Φ' is the space of tempered distributions on $\mathbb{R}$. The eigenkets $|a_R\rangle$ are in Φ'.

ignorant of which group element corresponds to this relation, he will describe the joint state of the reference token and system by the output of the encoding operation

$$\mathcal{E}_\tau : \ \mathcal{S}(\mathcal{H}_S) \rightarrow \mathcal{S}(\mathcal{H}_R \otimes \mathcal{H}_S)$$

$$\rho_S \mapsto \mathcal{E}_\tau[\rho_S] := \mathcal{G}_\tau[\rho_R \otimes \rho_S], \tag{7.9}$$

where the map $\mathcal{G}_\tau$ is a twirl over the finite interval $[-\tau, \tau] \subset G$,

$$\mathcal{G}_\tau [\rho_R \otimes \rho_S] := \frac{1}{2\tau} \int_{-\tau}^{\tau} dg \, \mathcal{U}_R(g)[\rho_R] \otimes \mathcal{U}_S(g)[\rho_S], \tag{7.10}$$

where dg is the Haar measure associated with G.

7.2.2   The Recovery Operation

As demonstrated by Bartlett et al. [3], Bob may perform a recovery operation $\mathcal{R}$ by first making a measurement of the reference token, followed by a reorientation of the system conditioned on the outcome of the measurement, and then discarding both the reference token and measurement result. We follow this procedure in constructing the recovery operation to be applied to the encoded state $\mathcal{E}_\tau(\rho_S)$.

Bob will make a measurement R of the reference token described by the POVM elements

$$R := \left\{ dg \, E(g), \ \forall g \in [-\tau, \tau] \subset G \right\} \cup \left\{ E_\tau \right\}, \tag{7.11}$$

where

$$E_\tau := I_R - \int_{-\tau}^{\tau} dg \, E(g), \tag{7.12}$$

$dg \, E(g)$ is the POVM element associated with outcome $g \in G$, and I_R is the identity operator on $\mathcal{H}_R$. We assume[3] that these POVM elements satisfy the covariance relation

$$\mathcal{U}_R(g')[E(g)] = E(g + g') \quad \forall g \in G. \tag{7.13}$$

---

[3]To the best of the author's knowledge the question of whether such a measurement exists for any G is an open problem, as suggested by the remarks in Sec. III.4.4 of Ref. [4]. However, it is suggested in this reference that it seems plausible that such a measurement can be constructed, although there does not seem to be an easy general procedure for its construction.

If the outcome of the measurement of R is $g \in [-\tau, \tau]$, associated with the POVM element $dg\, E(g)$, then Bob will reorient the system state ρ_S by implementing the unitary map $\mathcal{U}_S(g^{-1})$, which corresponds to the transformation of the reference token by an amount indicated by the measurement result (first term in Eq. (7.14)). If the outcome of the measurement is associated with the operator E_τ, Bob will do nothing (second term in Eq. (7.14)). After this measurement and reorientation, Bob will discard (trace out) the reference token and measurement result. This entire procedure will constitute the recovery operation $\mathcal{R}_\tau$.

The action of the recovery operation $\mathcal{R}_\tau$ on the encoded state $\mathcal{E}_\tau[\rho_S]$ is given by

$$
\begin{aligned}
\rho_S'(\tau) &= \mathcal{R}_\tau \circ \mathcal{E}_\tau[\rho_S] \\
&= \frac{1}{2\tau} \int_{-\tau}^{\tau} dg' \int_{-\tau}^{\tau} dg \,\, \mathrm{tr}\left(E(g')\mathcal{U}_R(g)[\rho_R]\right)\mathcal{U}_S(g'^{-1}) \circ \mathcal{U}_S(g)[\rho_S] \\
&\quad + \frac{1}{2\tau} \int_{-\tau}^{\tau} dg \,\, \mathrm{tr}\left(E_\tau\,\mathcal{U}_R(g)[\rho_R]\right)\mathcal{U}_S(g)[\rho_S].
\end{aligned}
\tag{7.14}
$$

7.2.3   Taking the Limit $\tau \to \infty$

The limit of Eq. (7.14) in which τ becomes infinite corresponds to the scenario in which Bob knows nothing about the orientation of his reference frame with respect to Alice's—the G-twirl appearing in the encoding map in Eq. (7.9) is an average over the entire group G.

As is clear from Eq. (7.12), in the limit $\tau \to \infty$ the operator E_τ vanishes, and thus the second term in Eq. (7.14) goes to zero. Taking this into account, the $\tau \to \infty$ limit of Eq. (7.14) is

$$
\rho_S' = \lim_{\tau \to \infty} \frac{1}{2\tau} \int_{-\tau}^{\tau} dg' \int_{-\tau}^{\tau} dg \,\, \mathrm{tr}\left(E(g' - g)\rho_R\right)\mathcal{U}_S(g - g')[\rho_S],
\tag{7.15}
$$

where we have used the covariance property of the POVM elements expressed in Eq. (7.13). Changing the integration variables to $u := g' - g$ and $v := g'$, the recovered state becomes

$$
\rho_S' = \lim_{\tau \to \infty} \frac{1}{2\tau} \int_{-\tau}^{\tau} dv \int_{v-\tau}^{v+\tau} du \,\, \mathrm{tr}\left(E(u)\rho_R\right)\mathcal{U}_S^\dagger(u)[\rho_S].
\tag{7.16}
$$

Denoting the antiderivative of the above integrand as

$$
F(x) := \int_0^x du \,\, \mathrm{tr}\left(E(u)\rho_R\right)\mathcal{U}_S^\dagger(u)[\rho_S],
\tag{7.17}
$$

equation (7.16) takes the form

$$\rho'_S = \lim_{\tau \to \infty} \frac{1}{2\tau} \int_{-\tau}^{\tau} dv \left(F(v + \tau) - F(v - \tau) \right). \tag{7.18}$$

Making the substitution $h := \tau + v$ and $h := \tau - v$ in the first and second terms, respectively, the recovered state simplifies to

$$\rho'_S = \lim_{\tau \to \infty} \frac{1}{2\tau} \int_0^{2\tau} dh \left(F(h) - F(-h) \right). \tag{7.19}$$

Taking the limit by applying l'Hôpital's rule [11] yields

$$\begin{aligned}
\rho'_S &= \frac{1}{2} \lim_{\tau \to \infty} \frac{\partial}{\partial \tau} \int_0^{2\tau} dh \left(F(h) - F(-h) \right) \\
&= \lim_{\tau \to \infty} \left(F(\tau) - F(-\tau) \right) \\
&= \int_G dg \ \mathrm{tr} \left(E(g)\rho_R \right) \mathcal{U}_S(g)[\rho_S],
\end{aligned} \tag{7.20}$$

where the integration is carried out over the entire group G.

This brings us to our main result: Even though the action of the G-twirl over a noncompact group yields non-normalizable states [12], the composition of the encoding operation, which makes use of the G-twirl, with the recovery operation applied to ρ_S results in a properly normalized state in $\mathcal{S}(\mathcal{H}_S)$. Explicitly

$$\rho'_S = \lim_{\tau \to \infty} \mathcal{R}_\tau \circ \mathcal{E}_\tau[\rho_S] = \int_G dg \ p(g) \mathcal{U}_S(g)[\rho_S] \in \mathcal{S}(\mathcal{H}_S), \tag{7.21}$$

where $p(g) := \mathrm{tr} \left(E(g)\rho_R \right)$ is a normalized probability distribution on G.

Equation (7.21) is identical to the expression for the composition of the recovery and encoding map defined for compact groups given in Eq. (7.4). From Eq. (7.21) we see that if $p(g)$ is highly peaked around the identity group element, then the only unitary that will contribute significantly is the identity operator, and the state recovered by Bob will be close to the state sent by Alice, $\rho'_S \approx \rho_S$. Thus, the success of the recovery operation, and consequently the quality of the reference token, can be quantified in terms of the width of $p(g)$, analogous to the compact case [3].

By expressing ρ_S in the basis furnished by the eigenkets of the generator A_S of the group G, we find the recovered state to be

$$\begin{aligned}
\rho'_S &= \int_G dg \ p(g) \int da_S da'_S \ \rho_S(a_S, a'_S) e^{iA_S g} |a_S\rangle\langle a'_S| e^{-iA_S g} \\
&= \int da_S da'_S \left[\int_G dg \ p(g) e^{ig(a_S - a'_S)} \right] \rho_S(a_S, a'_S) |a_S\rangle\langle a'_S| \\
&= \int da_S da'_S \ \tilde{p}(a_S - a'_S)\rho_S(a_S, a'_S) |a_S\rangle\langle a'_S|,
\end{aligned} \tag{7.22}$$

where in the last equality we have defined the Fourier transform of $p(g)$,

$$\tilde{p}(a_S - a'_S) := \int_G dg\, p(g)\, e^{ig(a_S - a'_S)}. \tag{7.23}$$

From the definition of the characteristic function $\tilde{p}(a_S - a'_S)$ above, we see that if $a_S = a'_S$, then $\tilde{p}(a_S - a'_S) = 1$, and consequently the diagonal elements of ρ_S are unaffected by the action of the communication channel $\lim_{\tau \to \infty} \mathcal{R}_\tau \circ \mathcal{E}_\tau$. On the other hand, since the characteristic function is bounded, $|\tilde{p}(a_S - a'_S)| \le 1$, when $a_S \ne a'_S$ the off-diagonal elements of ρ'_S are equal to those of ρ_S multiplied by a factor whose magnitude is less than or equal to unity. From this observation we see that the decoherence induced by $\lim_{\tau \to \infty} \mathcal{R}_\tau \circ \mathcal{E}_\tau$ occurs in the basis furnished by the eigenkets associated with the generator A_S of the group G.

To quantify the success of the recovery operation—how close the recovered state ρ'_S is to the initial state ρ_S—we will make use of the fidelity $F(\rho'_S, \rho_S)$ between the recovered state ρ'_S and the state $\rho_S = |\psi_S\rangle\langle\psi_S| \in \mathcal{S}(\mathcal{H}_R)$ that Alice sent, which we will take to be pure

$$|\psi_S\rangle = \int da_S\, \psi_S(a_S)\, |a_S\rangle, \tag{7.24}$$

where $\psi_S(a_S) := \langle a_S | e \rangle$. The fidelity $F(\rho'_S, \rho_S)$ is then given by

$$
\begin{aligned}
F(\rho'_S, \rho_S) &:= \langle\psi_S|\rho'_S|\psi_S\rangle \\
&= \int_G dg\, p(g)\, |\langle\psi_S|U_S(g)|\psi_S\rangle|^2 \\
&= \int da_S da'_S\, \tilde{p}(a_S - a'_S)\, |\psi_S(a_S)|^2\, |\psi_S(a'_S)|^2.
\end{aligned}
\tag{7.25}
$$

7.3 Reference Frames Associated with the Translation Group

We now examine the recovered state $\rho'_S = \lim_{\tau \to \infty} \mathcal{R}_\tau \circ \mathcal{E}_\tau[\rho_S]$ when the relevant reference frame is associated with the one-dimensional translation group.

Consider Alice and Bob being completely ignorant of the relation between the spatial origins of their laboratories, i.e. the relation between their positional reference frames. The group formed by all possible changes of Alice's reference frame is the one-dimensional translation group T_1. The unitary representation of the group element $g \in T_1$ on the system is $U_S(g) \in \mathcal{U}_S(\mathcal{H}_S)$ and on the reference token is $U_R(g) \in \mathcal{U}_R(\mathcal{H}_R)$. These representations are generated by their respective momentum operators $A_S = P_S$ and $A_R = P_R$.

Suppose as a token of Alice's reference frame she prepares the state $|e_\sigma\rangle \in \mathcal{H}_R \simeq L^2(\mathbb{R})$, which we take to be a Gaussian state

$$|e_\sigma\rangle = \frac{1}{\pi^{1/4}\sqrt{\sigma}} \int dx_R \, e^{-x_R^2/2\sigma^2} \, |x_R\rangle, \tag{7.26}$$

where we have expressed $|e_\sigma\rangle$ in the basis furnished by the eigenkets $|x_R\rangle$ of the position operator X_R on $\mathcal{H}_R$ and $\sigma > 0$ is the spread of this state with respect to this basis. Note that the different orientations of this token state $|e_\sigma(g)\rangle := U(g)|e_\sigma\rangle$ are orthogonal in the limit that σ vanishes, $\lim_{\sigma\to 0}\langle e_\sigma(g)|e_\sigma(g')\rangle = \delta_{g,g'}$, imitating a classical reference frame as discussed in the proceeding section. In this limit token states corresponding to different positional reference frames are completely distinguishable from each other.

We must now construct the recovery measurement R for which the associated set of POVM elements satisfy the covariance relation in Eq. (7.13) with respect to the translation group T_1. One such set is given by the PVM elements associated with the position operator X_R, namely, $E(x) := |x_R\rangle\langle x_R|$ for all $x \in \mathbb{R} \simeq T_1$, where $|x_R\rangle$ denotes the eigenket of X_R associated with the eigenvalue x_R. This follows from the fact that the position and momentum operators acting on $\mathcal{H}_R$ satisfy the canonical commutation relation $[X_R, P_R] = i$, which implies that P_R generates translations of the operator X_R, or equivalently $U_R(g)|x_R\rangle = |x_R + g\rangle$. However, there is a more general set of POVM elements corresponding to unsharp measurements of the position operator constructed by the convolution of $E(x)$ with some confidence measure μ,

$$E^\mu(x) := \int d\mu(q) \, E(x+q). \tag{7.27}$$

Direct substitution of $E^\mu(x)$ into Eq. (7.13) shows that indeed these unsharp POVM elements are covariant with respect to the translation group. In what follows we consider the family of unsharp POVM elements $E_\delta^\mu(x)$ defined by choosing a Gaussian measure parametrized by $\delta > 0$,

$$E_\delta^\mu(x) := \int dq \, \frac{e^{-q^2/\delta^2}}{\sqrt{\pi}\delta} E(x+q). \tag{7.28}$$

In the limit $\delta \to 0$, we have $E_\delta^\mu(x) \to E(x)$.

Given that Alice prepared the reference token in the state $\rho_R = |e_\sigma\rangle\langle e_\sigma| \in \mathcal{S}(\mathcal{H}_R)$, the probability distribution $p(g)$ appearing in Eq. (7.21) is

$$p(g) := \mathrm{tr}\left(E_\delta^\mu(g)\rho_R\right) = \frac{e^{-\frac{g^2}{\sigma^2+\delta^2}}}{\sqrt{\pi}\sqrt{\sigma^2 + \delta^2}}. \tag{7.29}$$

We note that $p(g)$ is peaked around $g = 0$ with a width of $\sqrt{\sigma^2 + \delta^2}$. From Eq. (7.21), and the discussion that immediately follows, we see that the parameter $\sqrt{\sigma^2 + \delta^2}$ determines the quality of the recovery operation: The smaller σ and δ are, the more peaked $p(g)$ is around the identity element and the closer Bob's recovered state will be to the state sent by Alice.

As a concrete example, suppose Alice wishes to send Bob the state $\rho_S = |\psi_S\rangle\langle\psi_S|$, where $|\psi_S\rangle \in \mathcal{H}_S \simeq L^2(\mathbb{R})$ is a Gaussian state

$$|\psi_S\rangle = \frac{1}{\pi^{1/4}\sqrt{\Delta}} \int dx_S\, e^{i\mu_p x} e^{-(x_S-\mu_x)^2/2\Delta^2} |x_S\rangle, \tag{7.30}$$

with Δ the width of the Gaussian state in the position basis $|x_s\rangle$ for $\mathcal{H}_S$, and μ_x and μ_p its average position and momentum, respectively. Using Eq. (7.25), the fidelity between ρ_S and the state recovered by Bob ρ'_S is

$$F(\rho'_S, \rho_S) = \frac{\Delta}{\sqrt{\Delta^2 + \frac{1}{2}\left(\sigma^2 + \delta^2\right)}}. \tag{7.31}$$

As might be expected, in the limit where σ and δ vanish the fidelity $F(\rho'_S, \rho_S)$ is equal to unity and the recovered state is exactly equal to the state Alice wished to send to Bob, $\rho'_S = \rho_S$. This limit corresponds different orientations of the reference token described by Eq. (7.8) being orthogonal, thus imitating a classical reference frame, and the measurement of the token's position being carried out perfectly.

From Eq. (7.31) we also observe that states less localized in the position basis (larger Δ) are better recovered by Bob, as illustrated in Fig. 7.2 in which the fidelity is plotted as a function of $\sqrt{\sigma^2 + \delta^2}$ for different Δ. Note that the expression for the fidelity is independent of μ_x and μ_p, implying that for Gaussian states the success of the recovery operation is independent of where the state is localized in phase space.

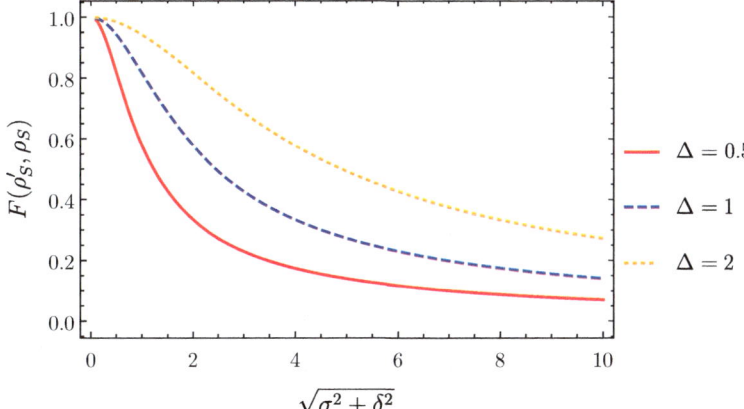

Fig. 7.2 The fidelity $F(\rho'_S, \rho_S)$ between the state sent by Alice ρ_S and the state recovered by Bob ρ'_S as a function of $\sqrt{\sigma^2 + \delta^2}$, where σ is the width of the reference token in position space and δ quantifies the accuracy of Bob's measurement of the reference token. It is seen that for a fixed $\sqrt{\sigma^2 + \delta^2}$, states less localized in the position basis (larger Δ) are better recovered by Bob

As a second example, suppose Alice prepares her token in a superposition of two Gaussian wave packets

$$|e\rangle = \frac{1}{\sqrt{N}}\Big(|\psi(\bar{x}, \bar{p}, \sigma)\rangle + |\psi(-\bar{x}, -\bar{p}, \sigma)\rangle \Big) \in \mathcal{H}_R, \qquad (7.32)$$

where N is an appropriate normalization constant and $|\psi(\bar{x}, \bar{p}, \sigma)\rangle$ denotes the state of a Gaussian wave packet of width σ in position space with average position and momentum $\bar{x}$ and $\bar{p}$, respectively. As they appear in Eq. (7.32), $\bar{x}$ and $\bar{p}$ quantify the size of the superposition in position and momentum space, respectively. Further, suppose that Bob is able to make a perfect measurement of the position of the reference token as described by the POVM elements $\lim_{\delta \to 0} E_\delta^\mu(x)$. And again, suppose Alice wishes to communicate the Gaussian state given in Eq. (7.30).

Given the above, the fidelity expressed in Eq. (7.25) yields

$$F(\rho'_S, \rho_S) = \beta \frac{e^{\beta^2 \bar{x}^2/\sigma^2} + e^{-\beta^2 \bar{p}^2 \sigma^2}}{e^{\bar{x}^2/\sigma^2} + e^{-\bar{p}^2 \sigma^2}}, \qquad (7.33)$$

where $\beta := \Delta/\sqrt{\Delta^2 + \sigma^2/2}$; note that $\beta \in (0, 1)$ and is equal to Eq. (7.31) when $\delta \to 0$. Further, β takes its maximum (minimum) value when $\Delta \gg \sigma$ ($\Delta \ll \sigma$). Also observe that the Fidelity in Eq. (7.33) is independent of μ_x and μ_p, implying that the success of the recovery operation is independent of where $|\psi_S\rangle$ is localized in phase space.

Observe that the Fidelity in Eq. (7.33) is a monotonically decreasing function of $\bar{x}$, which implies that Alice should prepare the size of the superposition in position space to be as small as possible (i.e. small $\bar{x}$) in order to maximize the fidelity. A second observation can be made by inspection of Fig. 7.3, which is a plot of both the maximum fidelity, $F_{\max} := \max\left[F(\rho'_S, \rho_S) \mid \bar{x}, \bar{p}, \sigma > 0 \right]$, and the value $\bar{p}_{\max}/\sigma$ which realizes this maximum as a function of the width Δ/σ of the state $|\psi_S\rangle$ Alice wishes to send to Bob; since the fidelity is monotonically decreasing in $\bar{x}\sigma$, this maximum occurs when $\bar{x}\sigma = 0$. From Fig. 7.3 we see that depending on the value of Δ/σ, Alice can adjust the state of the reference token by choosing the size of the superposition in momentum space $\bar{p}/\sigma$ so that the fidelity is maximized. That is, having the ability to create different sizes of superposition in momentum space can act as a resource to improve the communication channel specific to the state Alice wishes to send to Bob.

7.4 Conclusions and Outlook

We began by introducing a communication protocol between two parties, Alice and Bob, that do not share a reference frame associated with a compact group. Alice sends to Bob a token of her reference frame along with a system she wishes to

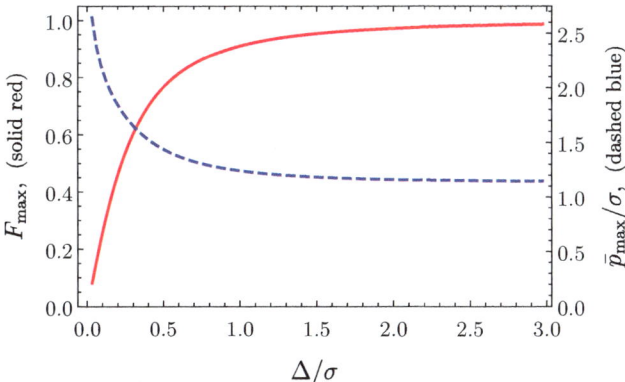

Fig. 7.3 For a reference token prepared in a superposition of two Gaussian states described by Eq. (7.32), the maximum fidelity $F_{\max} := \max\left[F(\rho'_S, \rho_S) \mid \bar{x}, \bar{p}, \sigma > 0\right]$ and the size of the superposition in momentum space $\bar{p}_{\max}/\sigma$ that realizes this maximum is plotted as a function of the width of the in position space Δ/σ of the state Alice sent to Bob as given in Eq. (7.30). For all values of Δ/σ the size of the superposition in momentum space which realizes the maximum fidelity is $\bar{x}_{\max}\sigma = 0$

communicate to him, and then Bob performs an appropriate recovery operation that enables him to recover a state close to the one Alice wished to communicate.

In Sect. 7.2 we demonstrated that this communication protocol can be applied when Alice's and Bob's reference frames are associated with a noncompact group, even though averaging states over the entire group leads to non-normalizable states. Furthermore, we demonstrated that this communication channel induces decoherence in the basis furnished by the eigenkets of the generator of the group. In Sect. 7.3 we applied this result to the study of communication between two parties who do not share a reference frame associated with the translation group. We introduced a sequence of Gaussian states $|e_\sigma\rangle$ of the reference token with spatial width σ and saw that in the limit $\sigma \to 0$, $|e_\sigma\rangle$ imitates a classical reference frame. This suggests that the parameter $1/\sigma$ acts as the size of the reference token, since as $1/\sigma$ becomes large the two parties are able to communicate perfectly (assuming Bob is able to measure the reference token perfectly, $\delta \to 0$). We also demonstrated that for finite-size reference tokens, i.e., when $1/\sigma$ is finite, states less localized in the position basis are better communicated to Bob and examined the case when the reference token is prepared in a superposition.

We note that the group of time translations generated by a Hamiltonian is a strongly continuous one-dimensional noncompact Lie group. Thus, provided a covariant measurement of the reference token corresponding to a time observable can be constructed [4], the above communication scheme can be employed. This will be fruitful for communication between parties who do not share a temporal reference frame, that is, their clocks are not synchronized. Furthermore, it will be interesting to see how the above construction can be applied to noncompact Lie groups of higher dimension, such as the Euclidean group in 2 and 3 dimensions, and ultimately the Poincaré group.

The intended application of the results in this chapter, as well as one of the primary motivations for this investigation, is to study the act of changing quantum reference frames.[4] Palmer et al. [10] have constructed an operational protocol for changing quantum reference frames associated with compact groups. They used the state $\mathcal{G}[\rho_A \otimes \rho_S]$ as a relational description of the state ρ_S with respect to a quantum reference frame ρ_A, and then considered the operation of changing the quantum reference frame from the state ρ_A to ρ_B. They found that this operation could not be done perfectly, and that the best one could do is

$$\mathcal{G}[\rho_A \otimes \rho_S] \rightarrow \mathcal{G}[\rho_B \otimes \rho_S'], \tag{7.34}$$

where $\rho_S' = \mathcal{R} \circ \mathcal{E}[\rho_S]$. In other words, one is not able to change quantum reference frames without affecting the state of the system described with respect to the reference frame—ρ_S changes to ρ_S' when the reference frame is changed. This results in a fundamental decoherence associated with the act of changing quantum reference frames. This decoherence is described by the composition of the encoding and recovery operations $\mathcal{R} \circ \mathcal{E}$ discussed in this chapter. Having generalized the operation $\mathcal{R} \circ \mathcal{E}$ to reference frames associated with noncompact groups, we hope to study the effect of changing quantum reference frames associated with the Galilean and Poincaré groups. Understanding the process of changing quantum reference frames is an essential step in the construction of a relational quantum theory, in which all objects, including reference frames, are treated quantum mechanically.

References

1. S.D. Bartlett, T. Rudolph, R.W. Spekkens, Classical and quantum communication without a shared reference frame. Phys. Rev. Lett. **91**, 027901 (2003)
2. S.D. Bartlett, T. Rudolph, R.W. Spekkens, Reference frames, superselection rules, and quantum information. Rev. Mod. Phys. **79**, 555 (2007)
3. S.D. Bartlett, T. Rudolph, R.W. Spekkens, P.S. Turner, Quantum communication using a bounded-size quantum reference frame. New J. Phys. **11**, 063013 (2009)
4. P. Busch, M. Grabowski, P.J. Lahti, *Operational Quantum Physics* (Springer, Berlin, 1997)
5. G. Chiribella, V. Giovannetti, L. Maccone, P. Perinotti, Teleportation transfers only speakable quantum information. Phys. Rev. A **86**, 010304(R) (2012)
6. R. de la Madrid, The role of the rigged Hilbert space in quantum mechanics. Eur. J. Phys. **26**, 287 (2005)
7. F. Giacomini, E. Castro-Ruiz, Č. Brukner, Quantum mechanics and the covariance of physical laws in quantum reference frames. Nat. Commun. **10**, 494 (2019)
8. P.A. Hoehn, A. Vanrietvelde, How to switch between relational quantum clocks (2018). arXiv:1810.04153 [gr-qc]

---

[4]See Refs. [7, 8, 14] for a different approach.

9. M.A. Nielsen, I.L. Chuang, *Quantum Computation and Quantum Information* (Cambridge University Press, Cambridge, 2010)
10. M.C. Palmer, F. Girelli, S.D. Bartlett, Changing quantum reference frames. Phys. Rev. A **89**, 052121 (2014)
11. W. Rudin, *Principles of Mathematical Analysis* (McGraw-Hill, New York, 1976)
12. A.R.H. Smith, M. Piani, R.B. Mann, Quantum reference frames associated with non-compact groups: the case of translations and boosts, and the role of mass. Phys. Rev. A **94**, 012333 (2016)
13. M.H. Stone, Linear transformations in hilbert space. III. Operational methods and group theory. Proc. Natl. Acad. Sci. U. S. A. **16**(2), 172–175 (1930)
14. A. Vanrietvelde, P.A. Hoehn, F. Giacomini, E. Castro-Ruiz, A change of perspective: switching quantum reference frames via a perspective-neutral framework (2018). arXiv:1809.00556 [quant-ph]
15. D. Verdon, J. Vicary, Perfect tight quantum teleportation without a shared reference frame. Phys. Rev. A **98**, 012306 (2018)
16. D. Verdon, J. Vicary, Quantum teleportation with infinite reference frame uncertainty and without prior alignment (2018). arXiv:1802.09040v2 [quant-ph]

Part III
Quantizing Time

Chapter 8
The Conditional Probability Interpretation of Time: The Case of Interacting Clocks

In quantum theory, time enters through its appearance as a classical parameter in the Schrödinger equation, as opposed to other physical quantities, such as position or momentum, which are associated with self-adjoint operators. Operationally, this time is what is measured by the clock on the wall of an experimenter's laboratory. This clock is a large classical object, not subject to quantum fluctuations, and does not interact with the system whose evolution it is tracking. Quantum theory describes the evolution of systems with respect to this clock. What changes when one tries to construct a quantum theory of spacetime?

The canonical quantization of gravity leads to the Wheeler-DeWitt equation: Physical states of the theory are annihilated by the Hamiltonian. In other words, the wave function of the universe—which includes the experimenter's clock, the system the experimenter is interested in, and everything else—is in the ground state of its Hamiltonian. Combined with the Schrödinger equation, the Wheeler-DeWitt equation dictates that the physical states of the theory do not evolve in time. How then do we explain the time evolution we see around us?

A necessary requirement for any quantum theory of gravity is to answer this question and explain how the familiar Schrödinger equation comes about from the Wheeler-DeWitt equation. The conditional probability interpretation of time offers an answer. As introduced by Page and Wootters [20–22, 25], the conditional probability interpretation defines the state of a system at a time t as a solution to the Wheeler-DeWitt equation conditioned on a subsystem of the universe, serving as a clock, to be in a state corresponding to the time t. Given an appropriate choice for the Hamiltonian of the universe and choice of clock, one finds this conditional state of the system satisfies the Schrödinger equation.

This interpretation of time was initially criticized by Kuchař [16, 21], who argued that it was unable to reproduce the correct two-time correlation functions, i.e. supposing a system was initially prepared in some state, what is the probability

© Springer Nature Switzerland AG 2019
A. R. H. Smith, *Detectors, Reference Frames, and Time*, Springer Theses,
https://doi.org/10.1007/978-3-030-11000-0_8

of finding the system in a different state at a later time? This criticism has since been overcome in two different ways: first, by correctly formulating the two-time correlation functions in terms of physical observables [7], and second, by modelling the measurement of the two-time correlation function as two successive von Neumann measurements [13].

Recently, the conditional probability interpretation of time has received considerable attention. Gambini et al. [9, 10, 12] have demonstrated that the conditional probability interpretation can result in a fundamental decoherence mechanism[1] and explored the consequences of this fact in relation to the black hole information loss problem [11]. Leon et al. [17] have shown that this interpretation overcomes Pauli's objection to constructing a time operator in quantum mechanics. Others have applied the formalism to a number of different systems and commented on various aspects of the proposal [2, 3, 18, 19].

The purpose of this chapter is to extend the conditional probability interpretation of time to take into account the possibility that the system being employed as a clock interacts with a system whose evolution the clock is tracking. As gravity couples everything, including clocks, this extension is necessary if the conditional probability interpretation of time is to be applied in a quantum gravitational setting. We find that taking into account a possible clock-system interaction within the conditional probability interpretation of time results in a time-nonlocal modification to the Schrödinger equation.

We begin in Sect. 8.1 by reviewing the extended phase space formulation of classical mechanics [15, 24] and its quantization. In Sect. 8.2 we introduce the conditional probability interpretation of time and its generalization to interacting clocks and systems, which results in a modified Schrödinger equation. We then show that this modified Schrödinger equation reduces to the familiar Schrödinger equation in an appropriate limit. We derive a series solution to this modified Schrödinger equation and demonstrate that the evolution it generates is time-nonlocal. We close this chapter in Sect. 8.3 with a summary of the results presented and comment on future directions of research.

8.1 The Hamiltonian Constraint in Classical and Quantum Mechanics

Both classical and quantum mechanics describe the evolution of the state, or equivalently the observables, of a physical system in time. However, general relativity demands that time itself be treated like any other physical system, that is, time should be treated dynamically. In this context, both classical and quantum

---

[1]The Schrödinger equation is replaced with a master equation, which induces decoherence.

mechanics describe relations between two physical systems: A clock, which indicates the time, and everything else. In this section we present a formulation of classical mechanics in which time is treated dynamically and on equal footing with the system whose evolution we are interested in. We then pass over to the corresponding quantum theory à la Dirac [6].

Consider a system S described by the action

$$S = \int_{t_1}^{t_2} dt\, L_S(q, q'),$$ (8.1)

where $L_S(q, q')$ is the Lagrangian associated with S, $q = q(t)$ denotes a set of generalized coordinates describing S, and $q' = q'(t)$ denotes the differentiation of these coordinates with respect to t.

Let us introduce an integration parameter τ and promote t to a dynamical variable $t(\tau)$, which we associate with the reading of a clock C. Through application of the chain rule the action above may be expressed as

$$S = \int_{\tau_1}^{\tau_2} d\tau\, \dot{t} L_S(q, \dot{q}/\dot{t}) = \int_{\tau_1}^{\tau_2} d\tau\, L(q, \dot{q}, \dot{t}),$$ (8.2)

where $L(q, \dot{q}, \dot{t}) := \dot{t} L_S(q, \dot{q}/\dot{t})$ is the Lagrangian describing both C and S and the dot denotes differentiation with respect to τ.

The Hamiltonian associated with $L(q, \dot{q}, \dot{t})$ is obtained by a Legendre transformation with respect to both $\dot{q}$ and $\dot{t}$

$$\tilde{H} = p_t \dot{t} + p_q \dot{q} - L(q, \dot{q}, \dot{t}) = \dot{t}\,(p_t + H_S),$$ (8.3)

where $H_S := p_q q' - L_S(q, q')$ is the Hamiltonian associated with $L_S(q, q')$ and we have used the fact that the momentum conjugate to q defined by $L(q, \dot{q}, \dot{t})$ is

$$p_q := \frac{\partial L(q, \dot{q}, \dot{t})}{\partial \dot{q}} = \dot{t}\frac{\partial L_S(q, \dot{q}/\dot{t})}{\partial (\dot{q}/\dot{t})}\frac{1}{\dot{t}} = \frac{\partial L_S(q, q')}{\partial q'},$$ (8.4)

which coincides with the momentum conjugate to q defined by $L_S(q, q')$. The momentum conjugate to t is

$$p_t := \frac{\partial L(q, \dot{q}, \dot{t})}{\partial \dot{t}} = L_S(q, q') - q' p_q = -H_S.$$ (8.5)

In light of Eq. (8.5), we see that term inside the brackets in Eq. (8.3) is constrained to vanish

$$H := p_t + H_S \approx 0. \tag{8.6}$$

We will refer to H as the total Hamiltonian as it describes both C and S.

It is natural to ask if the total Hamiltonian given in Eq. (8.6) is the most general possible? The answer is no. The total Hamiltonian can differ in two important ways:

1. An additional term $H_{int} = H_{int}(t, p_t, q, p_q)$ may be included in the total Hamiltonian, which couples C and S; this term will be referred to as the interaction Hamiltonian.
2. The momentum p_t may be replaced by a function of the conjugate variables associated with C, which we will refer to as the clock Hamiltonian and denote by $H_C = H_C(t, p_t)$. Note that in general C may be a composite system and have more than just one pair of conjugate variables.

Accounting for these generalizations, the most general total Hamiltonian is

$$H = H_C + H_S + \lambda H_{int} \approx 0, \tag{8.7}$$

where $\lambda \in \mathbb{R}$ is the strength of the interaction between C and S.

Motivating these generalization is general relativity, our best theory of time. The Hamiltonian formulation of general relativity does not admit a total Hamiltonian of the form given in Eq. (8.6). Total Hamiltonians that are linear in one of the conjugate momenta, like Eq. (8.6), indicate there is a preferred time variable in the theory [5], and this structure is not present in general relativity [15]. Further, gravity couples everything, including a clock and the system whose evolution it is tracking. Therefore, in a gravitational setting, we should expect an interaction Hamiltonian H_{int} to appear in the total Hamiltonian H coupling C and S.

We now wish to quantize the theory described by the total Hamiltonian given in Eq. (8.7). To do so, we follow the prescription given by Dirac [6]. We associate with C and S the Hilbert spaces $\mathcal{H}_C$ and $\mathcal{H}_S$, respectively. The total Hamiltonian H becomes an operator acting on the kinematical Hilbert space $\mathcal{H}_{kin} = \mathcal{H}_C \otimes \mathcal{H}_S$, and the constraint in Eq. (8.7) becomes

$$H |\psi\rangle\rangle = \left(H_C \otimes I_S + I_C \otimes H_S + \lambda H_{int} \right) |\psi\rangle\rangle = 0, \tag{8.8}$$

where I_C and I_S denote the identity operators on $\mathcal{H}_C$ and $\mathcal{H}_S$, respectively. The double ket notation is used to remind us that $|\psi\rangle\rangle$ is a state of both the clock and system. States $|\psi\rangle\rangle$ satisfying the constraint are in the physical Hilbert space $\mathcal{H}_{ph} \subset \mathcal{H}_{kin}$; states in the physical Hilbert space $|\psi\rangle\rangle \in \mathcal{H}_{ph}$ will be referred to as physical states. To completely specify the physical Hilbert space $\mathcal{H}_{ph}$ one must also choose an inner product on $\mathcal{H}_{ph}$, which we will do in the following section.

In general, the physical states evolve unitarily with respect to an external time, this evolution being generated by the total Hamiltonian. However, in totally constrained theories, such as the one defined by Eq. (8.8), the physical states are annihilated by the total Hamiltonian, $H | \psi \rangle \rangle = 0$, and therefore do not evolve with respect to any external time. The question then arises, how do we recover the dynamics we see around us from the frozen state $| \psi \rangle \rangle$? How does the Schrödinger equation come about from the constraint $H | \psi \rangle \rangle = 0$?

These questions constitute one aspect of the problem of time in quantum gravity [14, 16]. The conditional probability interpretation of time offers a way to reconcile the fact that the physical states are frozen with the time evolution we see around us. This is done by interpreting the outcome of a measurement of an observable on S at a specific time t, as a measurement of the physical state $| \psi \rangle \rangle$ conditioned on the clock being in a state corresponding to the time t.

8.2 The Conditional Probability Interpretation

We now introduce the conditional probability interpretation of time for theories described by the general total Hamiltonian given in Eq. (8.8). We will introduce the state of the system at a time t by conditioning a solution to the constraint, $| \psi \rangle \rangle$, on the clock being in a state corresponding to the time t. This state of the system will be seen to satisfy the Schrödinger equation in the limit where the interaction Hamiltonian H_{int} vanishes.

8.2.1 The Modified Schrödinger Equation

In the classical theory specified by the total Hamiltonian given in Eq. (8.6), time is defined operationally as the outcome of a measurement of the phase space variable t associated with a clock governed by the Hamiltonian $H_C = p_t$. In this case, the variable t is canonically conjugate to the clock Hamiltonian, $\{t, H_C\} = 1$.

The quantized version of this notion of time is to define time as a measurement of a time operator T on the clock Hilbert space $\mathcal{H}_C$ which is canonically conjugate to the clock Hamiltonian H_C

$$[T, H_C] = i. \tag{8.9}$$

In other words, states of the clock indicating different times correspond to eigenstates $|t\rangle$ of the time operator T, and the associated eigenvalue t is the time indicated by the clock. Employing the Baker-Campbell-Hausdorff formula, we see that as a consequence of the commutation relation in Eq. (8.9), H_C generates translations of T

$$e^{-iH_C s} T e^{iH_C s} = T - s I_C. \tag{8.10}$$

Resolving the identity on $\mathcal{H}_C$ as $I_C = \int dt\, |t\rangle\langle t|$ and making use of the spectral representation of the time operator $T = \int dt\, t\, |t\rangle\langle t|$, Eq. (8.10) may be expressed as

$$\int dt\, t e^{-iH_{CS}}\, |t\rangle\langle t|\, e^{iH_{CS}} = \int dt\, (t-s)\, |t\rangle\langle t| = \int dt\, t\, |t+s\rangle\langle t+s|\,, \qquad (8.11)$$

or

$$e^{-iH_{CS}}\, |t\rangle = |t+s\rangle\,, \qquad (8.12)$$

up to an overall phase. We will refer to the set of states $\{|t\rangle \mid \forall\, t\}$ as the clock states.

We now define the state of the system at time t as a solution to the constraint in Eq. (8.8) conditioned on the clock being in the state $|t\rangle$

$$|\psi_S(t)\rangle := \big((\langle t| \otimes I_S)\, |\psi\rangle\big)\,, \qquad (8.13)$$

where $|\psi_S(t)\rangle \in \mathcal{H}_S$. The state $|\psi_S(t)\rangle$ should be thought of as the time-dependent state of the system in the conventional formulation of quantum mechanics. With this definition of the system state, note that we may express the physical state $|\psi\rangle\rangle$ as

$$|\psi\rangle\rangle = \left(\int dt\, |t\rangle\langle t| \otimes I_S\right)|\psi\rangle\rangle = \int dt\, |t\rangle\, |\psi_S(t)\rangle\,. \qquad (8.14)$$

As mentioned above, we need to choose an inner product on the physical Hilbert space $\mathcal{H}_{ph}$. We will choose this inner product to be

$$\langle\langle \psi | \phi \rangle\rangle_{ph} := \langle\langle\psi|\big(|t\rangle\langle t| \otimes I_S\big)|\phi\rangle\rangle\,, \qquad (8.15)$$

for two states $|\psi\rangle\rangle$ and $|\phi\rangle\rangle$ in $\mathcal{H}_{ph}$. We will also demand that for states in $\mathcal{H}_{ph}$, this inner product is independent of the choice of t and that these states are normalized with respect to this inner product

$$\begin{aligned}
1 &= \langle\langle\psi|\psi\rangle\rangle_{ph} \\
&= \left[\int dt'\, \langle t'|\,\langle\psi_S(t')|\right]|t\rangle\langle t| \otimes I_S \left[\int dt''\, |t''\rangle\,|\psi_S(t'')\rangle\right] \\
&= \langle\psi_S(t)|\psi_S(t)\rangle\,. \qquad (8.16)
\end{aligned}$$

From the above equation, we see that this choice of inner product and normalization of the physical states ensure that the system state $|\psi_S(t)\rangle$ is properly normalized at all times.[2] Consequently, we can maintain the usual probabilistic interpretation of $|\psi_S(t)\rangle$.

Let us observe how $|\psi_S(t)\rangle$ changes with the parameter t labelling the clock states by acting on both sides of Eq. (8.13) with $i\,d/dt$:

$$
\begin{aligned}
i\frac{d}{dt}|\psi_S(t)\rangle &= i\frac{d}{dt}\big(\langle t|\otimes I_S\big)|\psi\rangle\rangle \\
&= -\langle t|\big(H_C\otimes I_S\big)|\psi\rangle\rangle \\
&= -\langle t|\big(H - I_C\otimes H_S - \lambda H_{int}\big)|\psi\rangle\rangle .
\end{aligned}
\tag{8.17}
$$

Using the fact that $H|\psi\rangle\rangle = 0$ we find $|\psi_S(t)\rangle$ satisfies

$$
i\frac{d}{dt}|\psi_S(t)\rangle = H_S|\psi_S(t)\rangle + \lambda\langle t|H_{int}|\psi\rangle\rangle .
\tag{8.18}
$$

Inserting a resolution of the identify on $\mathcal{H}_C$ in terms of the clock states $I_C = \int dt\,|t\rangle\langle t|$ between H_{int} and $|\psi\rangle\rangle$ in the second term of Eq. (8.18) and using the definition of the system state in Eq. (8.13), we find

$$
i\frac{d}{dt}|\psi_S(t)\rangle = H_S|\psi_S(t)\rangle + \lambda\int dt'\, A(t,t')|\psi_S(t')\rangle ,
\tag{8.19}
$$

where $A(t,t') := \langle t|H_{int}|t'\rangle$ is an operator acting on $\mathcal{H}_S$. We will refer to Eq. (8.19) as the modified Schrödinger equation. When the interaction Hamiltonian vanishes, $\lambda = 0$, the modified Schrödinger equation reduces to the usual Schrödinger equation.

The second term on the right-hand side of Eq. (8.19) is a linear integral operator on $\mathcal{H}_S$ with integration kernel $A(t',t)$; we will denote this integral operator as

$$
[H_A]|\psi_S(t)\rangle := \left[\int dt'\, A(t,t')\right]|\psi_S(t)\rangle = \int dt'\, A(t,t')\big|\psi_S(t')\rangle .
\tag{8.20}
$$

---

[2]We should emphasize this is a choice of inner product and normalization of the physical states, which may severely reduce the size of the physical Hilbert space. Furthermore, it may not be necessary to preserve the probabilistic interpretation of the system state. For example, the probabilistic interpretation may only be applicable in some limit, and it is the task of the physicist to explain how this limit comes about. To quote DeWitt on this point [4, 15]:

> ...one learns that time and probability are phenomenological concepts.

And Kiefer's clarification of DeWitt's statement [15]:

> The reference to probability refers to the 'Hilbert-space', problem, which is intimately connected with the 'problem of time'. If time is absent, the notion of a probability conserved in time does not make much sense; the traditional Hilbert-space structure was designed to implement the probability interpretation, and its fate in a timeless world thus remains open.

Note,[3] H_A is a self-adjoint operator if and only if $A(t, t') = A(t', t)^\dagger$. We see this is true from the definition of $A(t, t')$

$$A(t, t') := \langle t | H_{int} | t' \rangle = \big[\langle t' | H_{int} | t \rangle \big]^\dagger = A(t', t)^\dagger, \tag{8.21}$$

and therefore H_A is self-adjoint. If H_A is also a bounded operator,

$$\| H_A \| := \int dt dt' \, \| A(t, t') \|^2 < \infty, \tag{8.22}$$

then H_A is an integral operator of the Hilbert-Schmidt type [8, 26].

With the definition of H_A, let us write the modified Schrödinger equation in a more suggestive form

$$i \frac{d}{dt} | \psi_S(t) \rangle = \big[H_S + \lambda H_A \big] | \psi_S(t) \rangle . \tag{8.23}$$

Expressed this way, the modified Schrödinger equation can be seen as the ordinary Schrödinger equation with the system Hamiltonian H_S replaced with the self-adjoint integral operator $H_S + \lambda H_A$.

The evolution generated by the modified Schrödinger equation must preserve the norm of the system state as demanded by Eq. (8.16)

$$1 = \langle \psi_S(t) | \psi_S(t) \rangle \; \forall t \qquad \Rightarrow \qquad 0 = \frac{d}{dt} \langle \psi_S(t) | \psi_S(t) \rangle . \tag{8.24}$$

Evaluating $\frac{d}{dt} \langle \psi_S(t) | \psi_S(t) \rangle$ using Eq. (8.19) yields the condition

$$0 = \lambda \int dt' \left(\langle \psi_S(t) | A(t, t') | \psi_S(t') \rangle - \langle \psi_S(t) | A(t, t') | \psi_S(t') \rangle^* \right). \tag{8.25}$$

Since $A(t, t') := \langle t | H_{int} | t' \rangle$, Eq. (8.25) is a condition on the interaction Hamiltonian such that the modified Schrödinger equation is consistent with the normalization condition in Eq. (8.16).

We now show that Eq. (8.25) is satisfied for any choice of interaction Hamiltonian. Using Eq. (8.14) and the definition of the operator $A(t, t')$, the right-hand side of Eq. (8.25) can be expressed as

$$\langle\langle \psi | [\lambda H_{int}, | t \rangle \langle t | \otimes I_S] | \psi \rangle\rangle = \langle\langle \psi | [H - H_C \otimes I_S - I_C \otimes H_S, | t \rangle \langle t | \otimes I_S] | \psi \rangle\rangle$$

$$= - \langle\langle \psi | \big([H_C, | t \rangle \langle t |] \otimes I_S \big) | \psi \rangle\rangle$$

$$= -i \frac{d}{dt} \langle\langle \psi | \big(| t \rangle \langle t | \otimes I_S \big) | \psi \rangle\rangle$$

$$= -i \frac{d}{dt} \langle\langle \psi | \psi \rangle\rangle_{ph} , \tag{8.26}$$

---

[3] A proof of this can be found on pages 197–198 of [26].

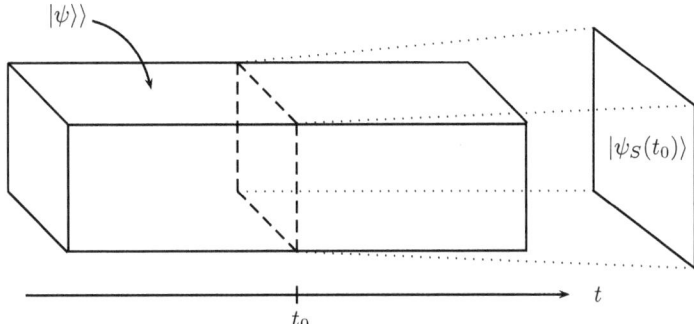

Fig. 8.1 The rectangular prism is a pictorial representation of the joint state of the clock and system $|\psi\rangle\rangle = \int dt\, |t\rangle\, |\psi_S(t)\rangle$. The horizontal axis represents the Hilbert space associated with the clock $\mathcal{H}_C$ and the directions orthogonal to the horizontal axis represent the Hilbert space of the system state $\mathcal{H}_S$. The system state $|\psi_S(t_0)\rangle$ at the time t_0 is obtained by conditioning $|\psi\rangle\rangle$ on the clock being in the state $|t_0\rangle$ and pictorially represented by a slice of the rectangular prism. Adapted from Giovannetti et al. [13]

which vanishes, since by construction $\langle\langle\psi|\psi\rangle\rangle_{ph} = 1$. Therefore, we conclude that the evolution generated by the modified Schrödinger equation preserves the norm of $|\psi_S(t)\rangle$.

From Eq. (8.14) we see that $|\psi\rangle\rangle$ describes an entangled state of the clock and system; see Fig. 8.1. This entanglement encodes the time evolution of the system state $|\psi_S(t)\rangle$ generated by the modified Schrödinger equation. This is somewhat analogous to the situation in general relativity. The state $|\psi\rangle\rangle$ is analogous to the four-dimensional spacetime metric—neither evolve with respect to an external time. However, one can foliate the four-dimensional spacetime by spacelike hypersurfaces (by choosing a clock), and then the four-dimensional metric encodes the evolution from one hypersurface to the next of the induced 3-metric and its conjugate momentum on these hypersurfaces. This evolution is analogous to the evolution of the system state $|\psi_S(t)\rangle$ governed by the modified Schrödinger equation.

If the interaction Hamiltonian happens to be of the form

$$\lambda H_{int} = \lambda \sum_i f_i(T) \otimes B_i, \tag{8.27}$$

where $f_i(T)$ is a function of the time operator T and B_i is a self-adjoint operator on $\mathcal{H}_S$, then the operator $A(t, t')$ is

$$\begin{aligned}
A(t, t') &= \langle t|H_{int}|t'\rangle \\
&= \sum_i \langle t|f_i(T)|t'\rangle\, B_i \\
&= \sum_i \langle t| \left(\int dt''\, f_i(t'')|t''\rangle\langle t''| \right) |t'\rangle\, B_i \\
&= \delta(t - t') \sum_i f_i(t)\, B_i.
\end{aligned} \tag{8.28}$$

Substituting Eq. (8.28) into the modified Schrödinger equation and simplifying, we find

$$i\frac{d}{dt}|\psi_S(t)\rangle = \left[H_S + \lambda \sum_i f_i(t)B_i\right]|\psi_S(t)\rangle, \tag{8.29}$$

which we recognize as the ordinary Schrödinger equation with H_S replaced with the time-dependent Hamiltonian

$$H_S(t) = H_S + \lambda \sum_i f_i(t)B_i. \tag{8.30}$$

8.2.2 Solving the Modified Schrödinger Equation

To further explore the consequences of the interaction Hamiltonian λH_{int} which couples the clock and system, we seek a series solution in the interaction strength λ to the modified Schrödinger equation.

Suppose the modified Schrödinger equation can be solved for $|\psi_S(t)\rangle$ in terms of a time evolution operator $V(t)$, so that the solution may be given as

$$|\psi_S(t)\rangle = V(t, t_0)|\psi_S(t_0)\rangle, \tag{8.31}$$

where $|\psi_S(t_0)\rangle \in \mathcal{H}_S$ is the state of the system at the time $t = t_0$ and $V(t_0, t_0) = I_S$. Suppose $V(t, t_0)$ may be expanded in powers of λ as

$$V(t, t_0) = \sum_{n=0}^{\infty} \lambda^n V_n(t, t_0). \tag{8.32}$$

Upon substituting $|\psi_S(t)\rangle = V(t, t_0)|\psi_S(t_0)\rangle$ into the modified Schrödinger equation and equating terms at equal order in λ, we find the operator $V_0(t, t_0)$ satisfies

$$i\frac{d}{dt}V_0(t, t_0) = H_S V_0(t, t_0) \quad \Rightarrow \quad V_0(t, t_0) = e^{-iH_S(t-t_0)}, \tag{8.33}$$

and we see that $V_0(t, t_0)$ is the usual Schrödinger time evolution operator $U(t, t_0) := e^{-iH_S(t-t_0)}$. The higher order operators $V_n(t, t_0)$ satisfy

$$i\frac{d}{dt}V_n(t, t_0) = H_S V_n(t, t_0) + \int dt'\, A(t, t')V_{n-1}(t', t_0), \tag{8.34}$$

the solution to which is the recurrence relation

$$V_n(t, t_0) = -iU(t, t_0)\int_{t_0}^{t} ds\, U(s, t_0)^{\dagger}\int du\, A(s, u)V_{n-1}(u, t_0). \tag{8.35}$$

Using Eqs. (8.33) and (8.35), the time evolution operator $V(t, t_0)$ may be expanded to leading order in λ as

$$V(t, t_0) = U(t, t_0) \left[I_S + (-i\lambda) \int_{t_0}^{t} ds \, U(s, t_0)^\dagger \int du \, A(s, u) U(u, t_0) + \mathcal{O}\left(\lambda^2\right) \right].$$
(8.36)

8.3 Summary

In this chapter we have generalized the conditional probability interpretation of time to account for an interaction between a clock and the system whose evolution it is tracking. This is a necessary consideration if the conditional probability interpretation is to be applied to any model of quantum gravity because gravity couples everything, including any clock and system. In the case of an interaction between the clock and system, we find the conditional state of the system $|\psi_S(t)\rangle$ satisfies a time-nonlocal modified Schrödinger equation.

We find that when the interaction Hamiltonian H_{int} is of the form given in Eq. (8.27), the modified Schrödinger equation becomes the ordinary Schrödinger equation with a time-dependent Hamiltonian dependent on H_{int}. In the limit when the interaction between the clock and system vanishes, $H_{int} = 0$, the modified Schrödinger equation reduces to the ordinary Schrödinger equation.

As it stands, the conditional probability interpretation of time does not specify a unique choice of clock states, and thus does not address the multiple choice problem [16]. In this chapter we chose clock states that are completely delocalized in the energy basis of the clock, that is, eigenstates of the operator canonically conjugate to the clock Hamiltonian. These clock states are maximally asymmetric under the action of the group generated by the clock Hamiltonian, which suggests that an appropriate figure of merit for choosing the clock states may come from the resource theory of asymmetry [1].

Future work will focus on realizing specific examples of the developed formalism. It should be noted that although the results presented were in the context of nonrelativistic quantum mechanics, in principle, there is nothing stopping the application of the conditional probability interpretation to relativistic quantum field theory and theories of quantum gravity.

Another avenue to explore is the possibility of replacing the infinite dimensional clock Hilbert space with a finite dimensional one. The canonical commutation relations between the clock Hamiltonian and time operator in Eq. (8.9) will no longer be satisfied. However, it is still possible to define a self-adjoint time operator that satisfies an approximate canonical commutation relation with the clock Hamiltonian [23]. It will be interesting to explore the role the dimension of the clock Hilbert space plays in a classical limit.

Another task will be to generalize the above formalism to mixed states. This will allow for the investigation of how an interaction between the system and clock affects the fundamental decoherence mechanism discussed by Gambini et al. [9–12].

References

1. S.D. Bartlett, T. Rudolph, R.W. Spekkens, Reference frames, superselection rules, and quantum information. Rev. Mod. Phys. **79**, 555 (2007)
2. K.L.H. Bryan, A.J.M. Medved, Realistic clocks for a universe without time (2017). arXiv:quant-ph/1706.02531
3. V. Corbin, N.J. Cornish, Semi-classical limit and minimum decoherence in the conditional probability interpretation of quantum mechanics. Found. Phys. **39**, 474 (2009)
4. B.S. DeWitt, The quantum and gravity: the Wheeler-DeWitt equation, in *The Eighth Marcel Grossmann Meeting*, ed. by T. Piran, R. Ruffini (World Scientific, Singapore, 1999), pp. 6–25
5. P.A.M. Dirac, Homogeneous variables in classical dynamics. Proc. Camb. Philos. Soc. **29**, 389 (1933)
6. P.A.M. Dirac, *Lectures on Quantum Mechanics* (Belfer Graduate School of Sciencem Yeshiva University, New York, 1964)
7. C.E. Dolby, The conditional probability interpretation of the Hamiltonian constraint (2004). arXiv:gr-qc/0406034
8. Hilbert-Schmidt operator. Encyclopedia of Mathematics. http://www.encyclopediaofmath.org/index.php?title=Hilbert-Schmidt_operator&oldid=22575
9. R. Gambini, R.A. Porto, J. Pullin, A relational solution to the problem of time in quantum mechanics and quantum gravity: a fundamental mechanism for quantum decoherence. New J. Phys. **6**, 45 (2004)
10. R. Gambini, R.A. Porto, J. Pullin, Loss of quantum coherence from discrete quantum gravity. Classical Quantum Gravity **21**, 51 (2004)
11. R. Gambini, R.A. Porto, J. Pullin, Realistic clocks, universal decoherence, and the block hole information paradox. Phys. Rev. Lett. **93**, 24041 (2004)
12. R. Gambini, R.A. Porto, S. Torterolo, J. Pullin, Conditional probabilities with Dirac observables and the problem of time in quantum gravity. Phys. Rev. D **79**, 041501 (2009)
13. V. Giovannetti, S. Lloyd, L. Maccone, Quantum time. Phys. Rev. D **79**, 945933 (2015)
14. C.J. Isham, Canonical quantum gravity and the problem of time, in *Integrable Systems, Quantum Groups, and Quantum Field Theories*, ed. by L.A. Ibort, M.A. Rodríguez. NATO ASI Series (Springer, Dordrecht, 1992)
15. C. Kiefer, *Quantum Gravity*, 3rd edn. (Oxford University Press, Oxford, 2012)
16. K.V. Kuchař, Time and interpretations of quantum gravity. Int. J. Mod. Phys. D **20**, 3 (2011)
17. J. Leon, L. Maccone, The Pauli objection. Found. Phys. **47**, 1597 (2017)
18. C. Marletto, V. Vedral, Evolution without evolution, and without ambiguities. Phys. Rev. D **95**, 043510 (2017)
19. E. Moreva, G. Brida, M. Gramegna, V. Giovannetti, L. Maccone, M. Genovese, Time from quantum entanglement: an experimental illustration. Phys. Rev. A **89**, 052122 (2014)
20. D.N. Page, Time as an inaccessible observable, NSF-ITP-89-18, 1989
21. D.N. Page, Clock time and entropy, in *Physical Origins of Time Asymmetry*, ed. by J.J. Halliwell, J. Pérez-Mercader, W.H. Zurek (Cambridge University Press, Cambridge, 1994)
22. D.N. Page, W.K. Wootters, Evolution without evolution: dynamics described by stationary observables. Phys. Rev. D **27**, 2885 (1983)
23. A. Peres, *Quantum Theory: Concepts and Methods* (Kluwer Academic Publishers, Dordrecht, 1995)
24. C. Rovelli, *Quantum Gravity* (Cambridge University Press, Cambridge, 2004)
25. W.K. Wootters, "Time" replaced by quantum correlations. Int. J. Theor. Phys. **23**, 701 (1984)
26. K. Yosida, *Functional Analysis*, 4th edn. (Springer, Berlin, 1980)

Epilogue: What Have We Learned?

Part I: Detectors in Curved Spacetimes

The way we learn about a system is by measuring it. This is true in quantum field theory on curved spacetime, in particular, with regard to learning about the entanglement structure of a field theory. The entanglement harvesting protocol is a measurement model that can be used to probe the entanglement structure of a quantum field theory. With this perspective, it was emphasized throughout that entanglement in quantum field theory depends on the measurement model employed. This is analogous to how the particle content of a field theory depends on the measuring apparatus, specifically on the motion of the observer carrying the apparatus.

In Chap. 3 we gave a physical motivation for a particular measurement model: the Unruh-DeWitt detector. We identified the POVM elements associated with this measurement model and derived the final state of two detectors moving along arbitrary timelike trajectories in spacetimes admitting a Wightman function. We quantified the amount of entanglement that appears in this state as a result of the detectors' local interaction with the field in terms of several measures of entanglement.

In Chap. 4 we applied these results to study how the entanglement structure of the Minkowski vacuum is affected by topological identifications. Why should we be interested in the global topology of spacetime? The reason is that the topology of our Universe may be non-trivial, in which case these considerations are bound to be important [8].

Chapter 5 investigated the behaviour of Unruh-DeWitt detectors operating in the exterior region of both the BTZ and $\mathbb{R}\mathbf{P}^2$ geon black holes. The response of a detector in the $\mathbb{R}\mathbf{P}^2$ geon spacetime was shown to be different than an identical detector in the BTZ spacetime, even though these spacetimes are locally indistinguishable from one another in the region in which the detectors are operating. We saw that the transition rate of a detector in the exterior of the $\mathbb{R}\mathbf{P}^2$ geon spacetime developed a time dependence as a result of the non-stationary features hidden behind

© Springer Nature Switzerland AG 2019
A. R. H. Smith, *Detectors, Reference Frames, and Time*, Springer Theses,
https://doi.org/10.1007/978-3-030-11000-0

its horizons. The point being that information about the difference between the global topology of the BTZ and $\mathbb{R}\mathbf{P}^2$ spacetimes is encoded in the vacuum state of a quantum field such that it is accessible by local measurements of the field.

We then applied the entanglement harvesting protocol developed in Chap. 3 to examine the entanglement structure of the Hartle-Hawking vacuum associated with the BTZ black hole, and how it depends on the parameters defining the spacetime. This served as an example of the entanglement harvesting protocol in a curved spacetime. We saw that as the detectors moved closer to the horizon, the entanglement that appears between them decreases. We gave an interpretation of this result in terms of the local Hawking temperature experienced by the detectors and red shift effects.

Part II: Quantum Reference Frames

Part II asked: What happens if we replace a classical reference frame with a quantum one? Our focus was on spatial reference frames associated with the translation group and inertial reference frames associated with the group of Galilean boosts. The motivation for studying these reference frames was that they are the simplest examples of reference frames associated with noncompact groups. If we are to apply the theory of quantum reference frames to relativistic scenarios where the relevant group is the Poincaré group, we must understand how the theory generalizes to reference frames associated with noncompact groups.

Chapter 6 demonstrated the failure of the G-twirl to define a properly normalized relational state when the twirling operation is carried out over a noncompact group. For the group of translations and Galilean boosts, we identified a relational state as the trace over the centre-of-mass degrees of freedom of a composite system and showed that this state appears naturally from averaging over this group. We then studied the informational properties of transforming between an external description of a composite system to a fully relational one.

Chapter 7 generalized the communication protocol of Bartlett et al. [5] to parties communicating without a shared classical reference frame associated with a one-dimensional noncompact Lie group. As an example of this protocol, we considered two parties employing different classical reference frames associated with a spatial origin, and quantified how well they can communicate as a function of the reference token prepared by the sender.

Part III: Quantizing Time

One could argue that the main difficulty, both conceptually and mathematically, in constructing a quantum theory of gravity is that by definition it requires a background independent quantization scheme. This is dramatically different than other theories of physics. We cannot consider matter moving around on a fixed background geometry and apply the usual quantization techniques; rather, we need to quantize matter and spacetime together. This strongly motivates a relational point of view.

The conceptual appeal of the conditional probability interpretation of time is that it is manifestly relational: everything is quantized, a clock is chosen as a subsystem of the universe, and the state of a system comprising everything else is defined relative to (conditioned on) the state of the clock. The time evolution of the system is encoded in the correlations of the entangled state of the clock and system satisfying the Wheeler-DeWitt equation. I find it satisfying that the relativity principle in the conditional probability interpretation is realized through entanglement.

Part III generalized the conditional probability interpretation to take into account the possibility of a coupling between the clock and the rest of the universe. We should expect such a coupling when the gravitational interaction between the clock and system is taken into account. We saw that what results is a time-nonlocal modification to Schrödinger equation. Future work will focus on exploring the full consequence of this fact and constructing explicit examples of the developed formalism.

In both classical and quantum mechanics, space and time are the entities we use to describe the dynamics of a given system. Quantum gravity is the quantization of space and time. This quantization will surely have an effect on the dynamics governing ordinary quantum theory in some limit.

Concluding Thought

Applying the theory of quantum information to situations at the boundary of relativity and quantum theory will certainly lead to new insights into the nature of our world. The hope is that this thesis takes a small step in that direction.

Appendix A
Derivation of Eq. (3.27)

The final state of two initially unexcited Unruh-Dewitt detectors interacting with the vacuum state of a scalar field is given by Eq. (3.24)

$$
\rho_{AB} = \begin{pmatrix} \rho_{11} & 0 & 0 & \rho_{14} \\ 0 & \rho_{22} & \rho_{23} & 0 \\ 0 & \rho_{23}^* & \rho_{33} & 0 \\ \rho_{14}^* & 0 & 0 & \rho_{44} \end{pmatrix},
\tag{A.1}
$$

where $\rho_{44} = E + \mathcal{O}(\lambda^6)$ and E is the leading order contribution given by

$$
\lambda^4 \int_{t>t'} dt\,dt' \int_{T>T'} dT\,dT'
$$

$$
\Big[\eta_A(t')\eta_B(t)\eta_B(T)\eta_A(T')e^{i[-\Omega_A \tau_A(t') - \Omega_B \tau_B(t) + \Omega_A \tau_A(T') + \Omega_B \tau_B(T)]}
$$

$$
\langle \phi_A(t')\phi_B(t)\phi_B(T)\phi_A(T') \rangle
$$

$$
+ \eta_B(t')\eta_A(t)\eta_B(T')\eta_A(T)e^{i[-\Omega_B \tau_B(t') - \Omega_A \tau_A(t) + \Omega_B \tau_B(T') + \Omega_A \tau_A(T)]}
$$

$$
\langle \phi_B(t')\phi_A(t)\phi_A(T)\phi_B(T') \rangle
$$

$$
+ \eta_A(t')\eta_B(t)\eta_B(T')\eta_A(T)e^{i[-\Omega_A \tau_A(t') - \Omega_B \tau_B(t) + \Omega_B \tau_B(T') + \Omega_A \tau_A(T)]}
$$

$$
\langle \phi_A(t')\phi_B(t)\phi_A(T)\phi_B(T') \rangle
$$

$$
+ \eta_B(t')\eta_A(t)\eta_A(T')\eta_B(T)e^{i[-\Omega_B \tau_B(t') - \Omega_A \tau_A(t) + \Omega_A \tau_A(T') + \Omega_B \tau_B(T)]}
$$

$$
\langle \phi_B(t')\phi_A(t)\phi_B(T)\phi_A(T') \rangle \Big].
\tag{A.2}
$$

© Springer Nature Switzerland AG 2019
A. R. H. Smith, *Detectors, Reference Frames, and Time*, Springer Theses,
https://doi.org/10.1007/978-3-030-11000-0

We wish to express the 4-point functions appearing above in terms of Wightman functions. Let us express each field in terms of its positive and negative frequency parts as $\phi_i = \phi_i^+ + \phi_i^-$, where

$$\phi_i^+ = \int d\mu(k)\, u_k(x_i) a_k \qquad \text{and} \qquad \phi_i^- = \int d\mu(k)\, u_k(x_i)^* a_k^\dagger, \qquad (A.3)$$

and note $\phi_i^+ |0\rangle = 0$ and $\langle 0| \phi_i^- = 0$; further

$$\phi_i^+ \phi_j^- = \phi_j^- \phi_i^+ + \left[\phi_i^+, \phi_j^-\right], \qquad (A.4)$$

and

$$\left[\phi_i^+, \phi_j^-\right] = \int d\mu(k) d\mu(p)\, u_k(x_i) u_p^*(x_j) \left[a_k, a_p^\dagger\right]$$

$$= \int d\mu(k)\, u_k(x_i) u_k^*(x_j) \mathbb{I}$$

$$= W(x_i, x_j) \mathbb{I}. \qquad (A.5)$$

Using these observations we can express an arbitrary 4-point function in terms of the Wightman function as

$$\langle \phi_1 \phi_2 \phi_3 \phi_4 \rangle = \langle \phi_1^+ \left(\phi_2^+ + \phi_2^-\right) \left(\phi_3^+ + \phi_3^-\right) \phi_4^- \rangle$$

$$= \langle \phi_1^+ \left(\phi_2^+ \phi_3^+ + \phi_2^+ \phi_3^- + \phi_2^- \phi_3^+ + \phi_2^- \phi_3^-\right) \phi_4^- \rangle$$

$$= \langle \phi_1^+ \phi_2^+ \phi_3^- \phi_4^- \rangle + \langle \phi_1^+ \phi_2^- \phi_3^+ \phi_4^- \rangle. \qquad (A.6)$$

In the above equation, the second term simplifies to

$$\langle \phi_1^+ \phi_2^- \phi_3^+ \phi_4^- \rangle = \langle \left(\phi_2^- \phi_1^+ + \left[\phi_1^+, \phi_2^-\right]\right) \left(\phi_4^- \phi_3^+ + \left[\phi_3^+, \phi_4^-\right]\right) \rangle$$

$$= \langle \left[\phi_1^+, \phi_2^-\right] \left[\phi_3^+, \phi_4^-\right] \rangle$$

$$= W(x_1, x_2) W(x_3, x_4), \qquad (A.7)$$

while the first term simplifies to

$$\langle \phi_1^+ \phi_2^+ \phi_3^- \phi_4^- \rangle = \langle \phi_1^+ \left(\phi_3^- \phi_2^+ + \left[\phi_2^+, \phi_3^-\right]\right) \phi_4^- \rangle$$

$$= \langle \phi_1^+ \phi_3^- \phi_2^+ \phi_4^- \rangle + \langle \phi_1^+ \left[\phi_2^+, \phi_3^-\right] \phi_4^- \rangle$$

$$= W(x_1, x_3) W(x_2, x_4) + W(x_2, x_3) \langle \phi_1^+ \phi_4^- \rangle$$

$$= W(x_1, x_3) W(x_2, x_4) + W(x_2, x_3) W(x_1, x_4). \qquad (A.8)$$

Therefore we may express the 4-point function as

$$\langle\phi_1\phi_2\phi_3\phi_4\rangle = W(x_1, x_2)W(x_3, x_4) + W(x_1, x_3)W(x_2, x_4)$$
$$+ W(x_2, x_3)W(x_1, x_4). \tag{A.9}$$

Applying Eq. (A.9) to each of the 4-point functions in Eq. (A.2), E may be expressed as

$$E = \lambda^4 \int_{t>t'} dtdt' \int_{T>T'} dTdT'$$
$$\left[E_{|X|^2}(t', t, T', T) + E_{|C|^2}(t', t, T', T) + E_{P_A P_B}(t', t, T', T) \right], \tag{A.10}$$

where

$$E_{|X|^2}(t', t, T', T)$$
$$:= \eta_A(t')\eta_B(t)\eta_B(T)\eta_A(T')e^{i\left[-\Omega_A\tau_A(t')-\Omega_B\tau_B(t)+\Omega_A\tau_A(T')+\Omega_B\tau_B(T)\right]}$$
$$\times W(x_A(t'), x_B(t))W(x_B(T), x_A(T'))$$
$$+ \eta_B(t')\eta_A(t)\eta_B(T')\eta_A(T)e^{i\left[-\Omega_B\tau_B(t')-\Omega_A\tau_A(t)+\Omega_B\tau_B(T')+\Omega_A\tau_A(T)\right]}$$
$$\times W(x_B(t'), x_A(t))W(x_A(T), x_B(T'))$$
$$+ \eta_A(t')\eta_B(t)\eta_B(T')\eta_A(T)e^{i\left[-\Omega_A\tau_A(t')-\Omega_B\tau_B(t)+\Omega_B\tau_B(T')+\Omega_A\tau_A(T)\right]}$$
$$\times W(x_A(t'), x_B(t))W(x_A(T), x_B(T'))$$
$$+ \eta_B(t')\eta_A(t)\eta_A(T')\eta_B(T)e^{i\left[-\Omega_B\tau_B(t')-\Omega_A\tau_A(t)+\Omega_A\tau_A(T')+\Omega_B\tau_B(T)\right]}$$
$$\times W(x_B(t'), x_A(t))W(x_B(T), x_A(T')), \tag{A.11}$$

$$E_{|C|^2}(t', t, T', T)$$
$$:= \eta_A(t')\eta_B(t)\eta_B(T)\eta_A(T')e^{i\left[-\Omega_A\tau_A(t')-\Omega_B\tau_B(t)+\Omega_A\tau_A(T')+\Omega_B\tau_B(T)\right]}$$
$$\times W(x_A(t'), x_B(T))W(x_B(t), x_A(T'))$$
$$+ \eta_B(t')\eta_A(t)\eta_B(T')\eta_A(T)e^{i\left[-\Omega_B\tau_B(t')-\Omega_A\tau_A(t)+\Omega_B\tau_B(T')+\Omega_A\tau_A(T)\right]}$$
$$\times W(x_B(t'), x_A(T))W(x_A(t), x_B(T'))$$
$$+ \eta_A(t')\eta_B(t)\eta_B(T')\eta_A(T)e^{i\left[-\Omega_A\tau_A(t')-\Omega_B\tau_B(t)+\Omega_B\tau_B(T')+\Omega_A\tau_A(T)\right]}$$
$$\times W(x_B(t), x_A(T))W(x_A(t'), x_B(T'))$$
$$+ \eta_B(t')\eta_A(t)\eta_A(T')\eta_B(T)e^{i\left[-\Omega_B\tau_B(t')-\Omega_A\tau_A(t)+\Omega_A\tau_A(T')+\Omega_B\tau_B(T)\right]}$$
$$\times W(x_A(t), x_B(T))W(x_B(t'), x_A(T')), \tag{A.12}$$

$$E_{P_A P_B}(t', t, T', T)$$

$$:= \eta_A(t')\eta_B(t)\eta_B(T)\eta_A(T')e^{i\left[-\Omega_A\tau_A(t')-\Omega_B\tau_B(t)+\Omega_A\tau_A(T')+\Omega_B\tau_B(T)\right]}$$

$$\times W(x_B(t), x_B(T))W(x_A(t'), x_A(T'))$$

$$+ \eta_B(t')\eta_A(t)\eta_B(T')\eta_A(T)e^{i\left[-\Omega_B\tau_B(t')-\Omega_A\tau_A(t)+\Omega_B\tau_B(T')+\Omega_A\tau_A(T)\right]}$$

$$\times W(x_A(t), x_A(T))W(x_B(t'), x_B(T'))$$

$$+ \eta_A(t')\eta_B(t)\eta_B(T')\eta_A(T)e^{i\left[-\Omega_A\tau_A(t')-\Omega_B\tau_B(t)+\Omega_B\tau_B(T')+\Omega_A\tau_A(T)\right]}$$

$$\times W(x_A(t'), x_A(T))W(x_B(t), x_B(T'))$$

$$+ \eta_B(t')\eta_A(t)\eta_A(T')\eta_B(T)e^{i\left[-\Omega_B\tau_B(t')-\Omega_A\tau_A(t)+\Omega_A\tau_A(T')+\Omega_B\tau_B(T)\right]}$$

$$\times W(x_B(t'), x_B(T))W(x_A(t), x_A(T')). \tag{A.13}$$

Observe that $E_{|X|^2}(t', t, T', T)$ factors as

$$E_{|X|^2}(t', t, T', T) = \left[\eta_A(t')\eta_B(t)e^{-i[\Omega_A\tau_A(t')+\Omega_B\tau_B(t)]}W(x_A(t'), x_B(t))\right.$$

$$\left. + \eta_B(t')\eta_A(t)e^{-i[\Omega_B\tau_B(t')+\Omega_A\tau_A(t)]}W(x_B(t'), x_A(t))\right]$$

$$\times \left[\eta_A(T')\eta_B(T)e^{i[\Omega_A\tau_A(T')+\Omega_B\tau_B(T)]}W(x_B(T), x_A(T'))\right.$$

$$\left. + \eta_B(T')\eta_A(T)e^{i[\Omega_B\tau_B(T')+\Omega_A\tau_A(T)]}W(x_A(T), x_B(T'))\right], \tag{A.14}$$

and thus the integration of $E_{|X|^2}(t', t, T', T)$ may be expressed as

$$\lambda^4 \int_{t>t'} dt dt' \int_{T>T'} dT dT'\, E_{|X|^2}(t', t, T', T)$$

$$= \left|\lambda^2 \int_{t>t'} dt dt' \left[\eta_A(t')\eta_B(t)e^{-i[\Omega_A\tau_A(t')+\Omega_B\tau_B(t)]}W(x_A(t'), x_B(t))\right.\right.$$

$$\left.\left. + \eta_B(t')\eta_A(t)e^{-i[\Omega_B\tau_B(t')+\Omega_A\tau_A(t)]}W(x_B(t'), x_A(t))\right]\right|^2, \tag{A.15}$$

which upon comparison with the definition X in Eq. (3.26), we see that

$$\lambda^4 \int_{t>t'} dt dt' \int_{T>T'} dT dT'\, E_{|X|^2}(t', t, T', T) = |X|^2. \tag{A.16}$$

Now observe from Eq. (A.12) that $E_{|C|^2}(t', t, T', T)$ has the following properties

$$E_{|C|^2}(t', t, T', T) = E_{|C|^2}(t, t', T', T) = E_{|C|^2}(t', t, T, T'), \qquad (A.17)$$

and that for a function $f(x, y)$, such that $f(x, y) = f(y, x)$, the following holds

$$\int_{x>y} dx dy\, f(x, y) = \int_{x<y} dx dy\, f(x, y) = \frac{1}{2} \int dx dy\, f(x, y). \qquad (A.18)$$

Using these observations, the integration of $E_{|C|^2}(t', t, T', T)$ specified by Eq. (A.10) is equivalent to

$$\int_{t>t'} dt dt' \int_{T>T'} dT dT'\, E_{|C|^2}(t', t, T', T)$$
$$= \frac{1}{4} \int dt dt' \int dT dT'\, E_{|C|^2}(t', t, T', T). \qquad (A.19)$$

By relabeling the integration variables for different terms in Eq. (A.12), it is seen that

$$\frac{1}{4} \int dt dt' \int dT dT'\, E_{|C|^2}(t', t, T', T)$$
$$= \left| \int dt dt'\, \eta_B(t)\eta_A(t') e^{i[\Omega_B \tau_B(t) - \Omega_A \tau_A(t')]} W(x_A(t'), x_B(t)) \right|^2. \qquad (A.20)$$

Then, using Eqs. (A.20) and (A.20) and comparing with the definition of C in Eq. (3.25), we find that

$$\int_{t>t'} dt dt' \int_{T>T'} dT dT'\, E_{|C|^2}(t', t, T', T) = |C|^2. \qquad (A.21)$$

By similar arguments leading to Eq. (A.21), one can show

$$\int_{t>t'} dt dt' \int_{T>T'} dT dT'\, E_{P_A P_B}(t', t, T', T) = P_A P_B, \qquad (A.22)$$

where P_A and P_B are defined by Eq. (3.9).

Therefore, using Eqs. (A.10), (A.16), (A.21), and (A.22), we find

$$E = |X|^2 + |C|^2 + P_A P_B, \qquad (A.23)$$

as stated in Eq. (3.27).

Appendix B
Mathematical Considerations

B.1 Quotient Spacetimes and the Method of Images

Some spacetime manifolds can be constructed as quotients by the action of a group on a different spacetime manifold. This quotient space construction is useful for understanding properties of these spacetimes that may have initially been constructed another way (for example, the BTZ black hole [3, 4]). In Part I of this thesis we exploited the quotient space structure of several spacetimes to easily construct Green's functions on them à la the method of images, which are necessary to describe the behaviour of detectors. In this appendix, the basic properties of quotient spacetimes are discussed and the method of images is introduced. This appendix follows closely the discussion in [6]; more details can be found there.

Consider a group G which acts on a spacetime manifold $\mathcal{M}$. The action of G on $\mathcal{M}$ is a map of sets $\mathcal{M} \times G \to \mathcal{M}$, denoted as $(p, g) \mapsto pg$, such that $pe = p$, with $e \in G$ being the identity element, and $(pg)g' = p(gg')$ for all $p \in \mathcal{M}$ and $g, g' \in G$. The G-orbit of $p \in \mathcal{M}$ is the set of points pg for all $g \in G$.

The space of all G orbits $\mathcal{M}/G$ is known as the quotient space. In general, a quotient space $\mathcal{M}/G$ will not be a manifold. The quotient space $\mathcal{M}/G$ is a manifold if and only if the group action of G is free and properly discontinuous.

A group action is free if the isotropy group $G_p := \{g \in G \mid pg = p\}$ is trivial at every point $p \in \mathcal{M}$, that is, for each $p \in \mathcal{M}$, $pg = p$ implies that $g = e$. A group action is properly discontinuous if:

1. Any two points $p, p' \in \mathcal{M}$ that do not lie on the same orbit have neighbourhoods U and U' such that $gU \cap U' = 0$ for all $g \in G$;
2. For each $p \in \mathcal{M}$, the isotropy group G_p is finite; and
3. Every point $p \in \mathcal{M}$ has a neighbourhood U such that $gU = U$ for $g \in G_p$, and $gU \cap U = 0$ for $g \notin G_p$.

The advantage of considering quantum field theories on quotient spacetimes is that Green's functions in the quotient spacetime $\mathcal{M}/G$ can be constructed via the

© Springer Nature Switzerland AG 2019

A. R. H. Smith, *Detectors, Reference Frames, and Time*, Springer Theses,
https://doi.org/10.1007/978-3-030-11000-0

method of images from the Green's function in the original spacetime $\mathcal{M}$. If we are given a Green's function $G_{\mathcal{M}}(x, x')$ on a spacetime $\mathcal{M}$, the corresponding Green's function $G_{\mathcal{M}/G}(x, x')$ on the quotient spacetime $\mathcal{M}/G$ is given by the image sum [1, 2]

$$G_{\mathcal{M}/G}(x, x') = \sum_n \eta^n G_{\mathcal{M}}(x, g^n x'), \tag{B.1}$$

where $g^n x'$ denotes the group action of the group element g^n on x' and $\eta = \{1, -1\}$ corresponding to untwisted and twisted fields, respectively.[1]

B.2   The Theory of Distributions

For convenience, we summarize here the basic properties of distributions we have used in this thesis and provide a proof of the identity used in obtaining Eq. (4.25).

Recall that the definition of a distribution G acting on a smooth test function $f(x)$ that tends to zero as $y \to \pm\infty$ is given by

$$\langle G, f \rangle := \text{PV} \int_{-\infty}^{\infty} dx \, g(x) f(x), \tag{B.2}$$

where the generalized function $g(x)$ defines the distribution G and PV specifies that the principle value of the integral should be taken. The derivative of a distribution is obtained from the above definition by integrating by parts to give

$$\langle G', f \rangle = -\langle G, f' \rangle. \tag{B.3}$$

The distribution $1/x$ acting on a test function $f(x)$ is defined as

$$\left\langle \frac{1}{x}, f(x) \right\rangle := \text{PV} \int_{-\infty}^{\infty} dx \, \frac{f(x)}{x}. \tag{B.4}$$

All the subsequent inverse power distributions $1/x^n$ are defined as distributional derivatives of $1/x$, hence

$$\left\langle \frac{1}{x^2}, f(x) \right\rangle = \left\langle \frac{1}{x}, f'(x) \right\rangle = \int_0^{\infty} dx \, \frac{f(x) + f(-x) - 2f(0)}{x^2}. \tag{B.5}$$

Equation (B.5) is used in arriving at the expression for transition probability of an Unruh-DeWitt detector in Minkowski space given in Eq. (4.27).

---

[1]More generally, $\eta = e^{i\delta}$. However, we will restrict ourselves to $\eta = \pm 1$, which corresponds to the case of untwisted and twisted fields.

Appendix C
Derivation of Eq. (5.27)

As derived in Sect. 5.3, in the sharp switching limit the transition rate of an Unruh-DeWitt detector outside the BTZ black hole turned on in the far past is given by

$$\frac{\dot{P}_{\mathrm{BTZ}}(\tau)}{\lambda^2} = \frac{1}{4} + \frac{1}{2\pi\sqrt{2}} \sum_{n=-\infty}^{\infty} \int_0^{\infty} d\tilde{s}$$

$$\mathrm{Re}\left[e^{-i\Omega\ell\tilde{s}} \left(\frac{1}{\sqrt{\sigma(x_D(\tau), \Gamma^n x_D(\tau - \ell\tilde{s}))}} \right. \right.$$

$$\left. \left. - \frac{\zeta}{\sqrt{\sigma(x_D(\tau), \Gamma^n x_D(\tau - \ell\tilde{s})) + 2}} \right) \right], \tag{C.1}$$

and for the detector trajectory specified in Eq. (5.24), we have

$$\sigma\left(x_D(\tau), \Gamma^n x_D(\tau - \ell\tilde{s})\right) = 2\frac{R^2 - r_h^2}{r_h^2}$$

$$\left[\frac{R^2}{R^2 - r_h^2} \sinh^2\left(n\pi\frac{r_h}{\ell}\right) - \sinh^2\left(\frac{r_h}{\sqrt{R^2 - r_h^2}}\frac{\tilde{s}}{2}\right) \right]. \tag{C.2}$$

For convenience, let us define

$$K_n := \frac{R^2}{R^2 - r_h^2} \sinh^2\left(n\pi\frac{r_h}{\ell}\right), \quad Q_n := K_n + \frac{r_h^2}{R^2 - r_h^2}, \quad \text{and}$$

$$\beta := 2\pi\sqrt{R^2 - r_h^2}/r_h, \tag{C.3}$$

© Springer Nature Switzerland AG 2019
A. R. H. Smith, *Detectors, Reference Frames, and Time*, Springer Theses,
https://doi.org/10.1007/978-3-030-11000-0

in terms of which the transition rate simplifies to

$$\frac{\dot{P}_{\text{BTZ}}(\tau)}{\lambda^2} = \frac{1}{4} + \frac{1}{2\beta} \times \sum_{n=-\infty}^{\infty} \int_0^\infty d\tilde{s}$$

$$\text{Re}\left[e^{-i\Omega\ell\tilde{s}} \left(\frac{1}{\sqrt{K_n - \sinh^2\left(\frac{\pi}{\beta}\tilde{s}\right)}} - \frac{\zeta}{\sqrt{Q_n - \sinh^2\left(\frac{\pi}{\beta}\tilde{s}\right)}} \right) \right].$$

$$(C.4)$$

We now focus on evaluating the integrals appearing above, which are of the form

$$I(a, P) := \text{Re} \int_0^\infty dx \, \frac{e^{-iax}}{\sqrt{P - \sinh^2 x}}. \tag{C.5}$$

Hodgkinson and Louko [7] evaluated $I(a, p)$, and we summarize their approach below.

We first examine the case when $P = 0$, in which Eq. (C.5) simplifies to

$$I(a, 0) = -\int_0^\infty dx \, \frac{\sin ax}{\sinh x} = -\frac{\pi}{2} \tanh \frac{\pi}{2} a, \tag{C.6}$$

where the last equality was obtained using Mathematica. Note that Eq. (C.6) may be expressed as

$$I(a, 0) = -\frac{\pi}{2} + e^{-\pi a/2} \, \text{Re} \int_0^\infty dy \frac{e^{-iay}}{\cosh y}, \tag{C.7}$$

which may be confirmed by directly evaluating the above integral and comparing with Eq. (C.6).

Supposing that $P > 0$, we may express Eq. (C.5) as a contour integral

$$I(a, P) = \text{Re} \int_{C_1} dz \, \frac{e^{-iaz}}{\sqrt{P - \sinh^2 z}}, \tag{C.8}$$

where C_1 is the contour from $z = 0$ to $z = \infty$ along the positive real axis with a dip into the lower half plane around the branch point $z = \arcsin\sqrt{P}$. The contour C_1 may be deformed into the union of the contours C_2 and C_3, where C_2 runs form $z = 0$ to $z = -i\pi/2$ along the negative imaginary axis and C_3 in the half line $z = y - i\pi/2$ for $y \in (0, \infty)$. As the integrand has no singularities for $\text{Im}\, z \in [-\pi/2, 0)$ and vanishes as $\text{Re}[z] \to \infty$ for $\text{Im}[z] < 0$, the deformation of C_1 into $C_2 \cup C_3$ does not change the value of the integral.

With these choices of contours, Eq. (C.8) becomes

$$
\begin{aligned}
I(a, P) &= \mathrm{Re} \int_{C_2} dz \, \frac{e^{-iaz}}{\sqrt{P - \sinh^2 z}} + \mathrm{Re} \int_{C_3} dz \, \frac{e^{-iaz}}{\sqrt{P - \sinh^2 z}} \\
&= \mathrm{Re} \int_0^{-\pi/2} dz \, \frac{i e^{az}}{\sqrt{P + \sin^2 z}} + \mathrm{Re} \int_0^{\infty} dy \, \frac{e^{-ia(y - i\pi/2)}}{\sqrt{P - \sinh^2(y - i\pi/2)}} \\
&= e^{-a\pi/2} \, \mathrm{Re} \int_0^{\infty} dy \, \frac{e^{-iay}}{\sqrt{P + \cosh^2 y}}.
\end{aligned}
\tag{C.9}
$$

Applying Eqs. (C.6) and (C.9) to the transition rate in Eq. (C.4) yields

$$
\dot{P}_{BTZ} = \frac{\lambda^2}{2\pi} e^{-\beta\Omega\ell/2} \sum_{n=-\infty}^{\infty} \int_0^{\infty} dy \cos(y\beta\Omega\ell/\pi)
$$

$$
\left[\frac{1}{\sqrt{K_n + \cosh^2 y}} - \frac{\zeta}{\sqrt{Q_n + \cosh^2 y}} \right],
\tag{C.10}
$$

as stated in Eq. (5.27).

References

1. R. Banach, J.S. Dowker, Automorphic field theory-some mathematical issues. J. Phys. A **12**, 2527 (1979)
2. R. Banach, J.S. Dowker, The vacuum stress tensor for automorphic fields on some flat space-times. J. Phys. A **12**, 2545 (1979)
3. M. Bañados, C. Teitelboim, J. Zanelli, The black hole in three-dimensional spacetime. Phys. Rev. Lett. **69**, 1849 (1992)
4. M. Bañados, M. Henneaux, C. Teitelboim, J. Zanelli, Geometry of the 2+1 black hole. Phys. Rev. D **48**, 1506 (1993)
5. S.D. Bartlett, T. Rudolph, R.W. Spekkens, P.S. Turner, Quantum communication using a bounded-size quantum reference frame. New J. Phys. **11**, 063013 (2009)
6. S. Carlip, The (2+1)-dimensional black hole. Classical Quantum Gravity **12**, 2853 (1995)
7. L. Hodgkinson, J. Louko, Static, stationary and inertial Unruh-DeWitt detectors on the BTZ black hole. Phys. Rev. D **86**, 064031 (2012)
8. R.D. Sorkin, Introduction to topological geons, in *Topological Properties and Global Structure of Space-Time*, ed. by P.G. Bergmann, V.D. Sabbata. NATO ASI Series (Plenum, New York, 1986), pp. 249–270

Index

© Springer Nature Switzerland AG 2019
A. R. H. Smith, *Detectors, Reference Frames, and Time*, Springer Theses,
https://doi.org/10.1007/978-3-030-11000-0

Printed by Printforce, the Netherlands